W9-AAE-547

KINGPIN

KINGPIN

HOW ONE HACKER TOOK OVER THE BILLION-DOLLAR CYBERCRIME UNDERGROUND

KEVIN POULSEN

Crown Publishers

New York

Published in the United States by Crown Publishers,
an imprint of the Crown Publishing Group,
a division of Random House, Inc., New York.
www.crownpublishing.com

Library of Congress Cataloging-in-Publication Data
Poulsen, Kevin, 1965–
Kingpin / Kevin Poulsen.—1st ed.
p. cm.
1. Butler, Max. 2. Computer crimes—United States—Case studies.
3. Computer hackers—United States—Case studies. 4. Commercial criminals—
United States—Case studies. I. Title.
HV6773.2.P68 2010
364.16'8092—dc22 2010027952

ISBN 978-0-307-58868-5
eISBN 978-0-307-58870-8

Printed in the United States of America

Book design by Leonard W. Henderson
Jacket design by Chris Sergio
Jacket photographs © Jonathan Kitchen/Photographer's Choice

1 3 5 7 9 10 8 6 4 2

First Edition

For Lauren,
my unindicted coconspirator in life

CONTENTS

COPS AND CARDERS

Max Vision, born Max Butler. Ran Carders Market under the handle Iceman. Also known as Ghost23, Generous, Digits, Aphex, and the Whiz.

Christopher Aragon, aka Easylivin', Karma, and the Dude. Max's partner on Carders Market, who ran a lucrative credit card counterfeiting ring fueled by Max's stolen data.

Script. A Ukrainian seller of stolen credit card data and founder of CarderPlanet, the first carder forum.

King Arthur. The Eastern European phisher and ATM cashout king who took over CarderPlanet from Script.

Maksik. The Ukrainian carder Maksym Yastremski, who replaced Script as the underground's top vendor of stolen credit card data.

Albert Gonzalez, aka Cumbajohnny and SoupNazi. An administrator on Shadowcrew, the largest crime site on the Web until the Secret Service took it down.

David Thomas, aka El Mariachi. A veteran scammer who ran a carding forum called the Grifters as an intelligence-gathering operation for the FBI.

John Giannone, aka Zebra, Enhance, MarkRich, and the Kid. A young carder from Long Island who worked with Max online and with Chris Aragon in real life.

J. Keith Mularski, aka Master Splyntr, Pavel Kaminski. The Pittsburgh-based FBI agent who took over DarkMarket in a high-stakes undercover operation.

Greg Crabb. A U.S. postal inspector, and Keith Mularski's mentor, who spent years tracking the underground's elusive international leaders.

Brett Johnson, aka Gollumfun. A Shadowcrew founder who went on to serve as an administrator on Carders Market.

Tea, aka Alenka. Tsengeltsetseg Tsetsendelger, a Mongolian immigrant who helped run Carders Market from a safe house in Orange County.

JiLsi. Renukanth Subramaniam, the Sri Lankan–born British citizen who founded DarkMarket.

Matrix001. Markus Kellerer, a German DarkMarket administrator.

Silo. Lloyd Liske, a Canadian hacker who became an informant for the Vancouver police.

Th3C0rrupted0ne. A former drug dealer and recreational hacker who served as an administrator on Carders Market.

PROLOGUE

The taxi idled in front of a convenience store in downtown San Francisco while Max Vision paid the driver and unfolded his six-foot-five frame from the back of the car, his thick brown hair pulled into a sleek ponytail. He stepped into the store and waited for the cab to disappear down the street before emerging for the two-block walk to his safe house.

Around him, tiny shops and newsstands awakened under the overcast sky, and suited workers filed into the office towers looming above. Max was going to work too, but his job wouldn't have him home after nine hours for a good night's sleep. He'd be cloistered for days this time. Once he put his plan into motion, there'd be no going home. No slipping out for a bite of dinner. No date night at the multiplex. Nothing until he was done.

This was the day he was declaring war.

His long gait took him to the Post Street Towers, from the street a five-by-fourteen grid of identical bay windows, trim painted the color of the Golden Gate Bridge. He'd been coming to this apartment complex for months, doing his best to blend in with the exchange students drawn by short leases and reasonable rents. Nobody knew his name—not his real one anyway. And nobody knew his past.

Here, he wasn't Max Butler, the small-town troublemaker driven by obsession to a moment of life-changing violence, and he wasn't Max Vision, the self-named computer security expert paid one hundred dollars an hour to harden the networks of Silicon Valley companies. As he rode up the apartment building elevator, Max became someone else: "Iceman"—a rising leader in a criminal economy responsible for billions of dollars in thefts from American companies and consumers.

And Iceman was fed up.

For months, he'd been popping merchants around the country, prying out piles of credit card numbers that should have been worth hundreds of thousands on the black market. But the market was broken. Two years earlier Secret Service agents had driven a virtual bulldozer through the computer underworld's largest gathering spot, arresting the ringleaders at gunpoint and sending the rest scurrying into chat rooms and small-time Web forums—all riddled with security holes and crawling with feds and snitches. It was a mess.

Whether they knew it or not, the underworld needed a strong leader to unify them. To bring order.

Off the elevator, Max idled in the hallway to check for a tail, then walked to his apartment door and entered the oppressive warmth of the rented studio. Heat was the biggest problem with the safe house. The servers and laptops crammed into the space produced a swelter that pulsed through the room. He'd brought in fans over the summer, but they provided scant relief and lofted the electric bill so high that the apartment manager suspected him of running a hydroponic dope farm. But it was just the machines, entwined in a web of cables, the most important snaking to a giant parabolic antenna aimed out the window like a sniper rifle.

Shrugging off his discomfort, Max sat at his keyboard and trained a bead on the Web forums where computer criminals gathered—virtual cantinas with names like DarkMarket and TalkCash. For two days, he hacked, his fingers flying at preternatural speed as he breached the sites' defenses, stealing their content, log-ins, passwords, and e-mail addresses. When he tired, he crashed out on the apartment's foldaway bed for an hour or two, then returned bleary-eyed to his work.

He finished with a few keystrokes that wiped out the sites' databases with the ease of an arsonist flicking a match. On August 16, 2006, he dispatched an unapologetic mass e-mail to the denizens of the sites he'd destroyed: They were all now members of Iceman's own Cardersmarket

.com, suddenly the largest criminal marketplace in the world, six thousand users strong and the only game in town.

With one stroke, Max had undermined years of careful law enforcement work and revitalized a billion-dollar criminal underworld.

In Russia and Ukraine, Turkey and Great Britain, and in apartments, offices, and houses across America, criminals would awaken to the announcement of the underground's first hostile takeover. Some of them kept guns in their nightstands to protect their millions in stolen loot, but they couldn't protect themselves from this. FBI and Secret Service agents who'd spent months or years infiltrating the now-destroyed underground forums would read the message with equal dismay, and for a moment, all of them—hacking masterminds, thuggish Russian mobsters, masters of fake identities, and the cops sworn to catch them—would be unified by a single thought.

Who is Iceman?

KINGPIN

1

The Key

As soon as the pickup truck rolled up to the curb, the teenage computer geeks squatting on the sidewalk knew there'd be trouble. "Fucking wavers!" one of the cowboys called out the window. A beer bottle flew from the truck and crashed on the pavement. The geeks, who'd left the club to talk away from the din of music, had seen it all before. In Boise in 1988, being caught in public without a wide belt buckle and a cowboy hat was a bottlin' offense.

Then one of the geeks did something the cowboys weren't expecting: He stood up. Tall and broad shouldered, Max Butler cut a quietly imposing figure that was enhanced by his haircut, a spiky punk-rock brush that added three inches to his height. "Waver?" Max asked calmly, feigning ignorance of the Boise slang for New Wave music fans and other freaks. "What's that?" The two cowboys blustered and swore, then finally drove away with a screech of tires and the waving of mud flaps.

Since they met one another in junior high, Max had become the unofficial bodyguard in the klatch of fellow computer nerds in Meridian, Idaho, a bedroom community then separated from Boise by eight miles of patchy farmland. The town fathers had named Meridian a century earlier for its placement directly on the Boise Meridian, one of the thirty-seven invisible north-south lines that form the Y-axes in America's land survey system. But that was probably the only thing geeky about the town, where the high school rodeo team got all the girls.

Max's parents had married young, and they'd moved to Idaho from Phoenix when he was an infant. In some ways, Max combined their best qualities: Robert Butler was a Vietnam veteran and enthusiastic technology buff who ran a computer store in Boise. Natalie Skorupsky was the daughter of Ukrainian immigrants—a humanist and a peacenik, she liked to relax in front of the Weather Channel and nature documentaries. Max inherited his mother's clean-living values, eschewing red meat, cigarettes, and alcohol and drugs, except for an ill-fated experiment with chewing tobacco. From his father, Max acquired a deep passion for computers. He grew up surrounded by exotic machines, from giant business computers that could double as an office desk to the first suitcase-sized "portable" IBM compatibles. Max was allowed to play with them freely. He started programming in BASIC at the age of eight.

But Max's equilibrium disappeared when his parents divorced in his fourteenth year. His father wound up in Boise, while Max lived in Meridian with his mother and his younger sister, Lisa. The divorce devastated the teenager and seemed to reduce him to two modes of operation: relaxed, and full-bore insane. When his manic side flared, the world was too slow to keep up; his brain moved at light speed and focused like a laser on whatever task was before him. After he got his driver's license, he drove his silver Nissan like the accelerator was a toggle switch, speeding from stop sign to stop sign, wearing lab goggles like a mad scientist conducting an experiment in Newtonian physics.

As Max protected his friends, they tried to protect Max from himself. His best buddy, a genial kid named Tim Spencer, found Max's world exciting but was constantly reining in his friend's impetuousness. One day he emerged from his home to find Max standing over an elaborate geometric pattern burning in the lawn. Max had found a canister of gasoline nearby. "Max, this is our house!" Tim shouted. Max sputtered apologies as the pair stamped out the blaze.

. . .

It was Max's impulsive side that made his friends resolve not to tell him about the key.

The Meridian geeks had found the key ring in an unlocked desk at the back of the chemistry lab. For a time, they just watched it, sliding open the desk drawer when the lab instructor wasn't around and checking to see if it was still there. Finally, they swiped it, smuggled it from the lab, and discreetly began testing its keys against various locks on the Meridian High campus. That was how they discovered that one of the keys was a master key to the school; it opened the front door and every door behind it.

Four copies were made, one for each of them: Tim, Seth, Luke, and John. The key ring was returned to the darkness of the chem lab desk after being carefully wiped down for fingerprints. They all agreed that Max must not know. A master key to the high school is a very special talisman that must be wielded with great care—not squandered on foolishness. So the juniors vowed to save the key for an epic senior-year prank. They would sneak into the school and hijack the PA system, blaring music into every classroom. Until that day, the four keys would stay in hiding, a burden borne in silence by the four of them.

Nobody liked keeping secrets from Max, but they could see that he was already on a collision course with the school's administrators. Max scoffed at the curriculum, and while instructors droned on about history or sketched equations on the blackboard, Max would sit at his desk thumbing through computer printouts from dial-up bulletin board systems and the pre-Web Internet. His favorite read was an online hacker newsletter called *Phrack,* a product of the late-1980s hacking scene. In its plain, unadorned text, Max could follow the exploits of editors Taran King and Knight Lightning, and contributors like Phone Phanatic, Crimson Death, and Sir Hackalot.

The first generation to come of age in the home computing era was tasting the power at its fingertips, and *Phrack* was a jolt of subversive, electric information from a world far beyond Meridian's sleepy borders. A typical issue was packed with tutorials on packet-switched networks like Telenet and Tymnet, guides to telephone-company computers like COSMOS, and inside looks at large-scale operating systems powering mainframe and mini-computers in air-conditioned equipment rooms around the globe.

Phrack also diligently tracked news reports from the frontier battleground between hackers and their opponents in state and federal law enforcement, who were just beginning to meet the challenges posed by recreational hackers. In July 1989, a Cornell graduate student named Robert T. Morris Jr. was charged under a brand-new federal computer crime law after he launched the first Internet worm—a virus that spread to six thousand computers, clogging network bandwidth and dragging systems to a halt. The same year, in California, a young Kevin Mitnick picked up his second hacking arrest and received one year in prison—a startlingly harsh sentence at the time.

Max became "Lord Max" on the Boise bulletin board systems and delved into phone phreaking—a hacking tradition dating to the 1970s. When he used his Commodore 64 modem to scan for free long-distance codes, he had his first run-in with the federal government: A Secret Service agent from the Boise field office visited Max at school and confronted him with the evidence of his phreaking. Because he was a juvenile, he wasn't charged. But the agent warned Max to change course before he got in real trouble.

Max promised he'd learned his lesson.

Then the unthinkable happened. Max noticed an odd shape on John's key ring and asked what it was. John confessed the truth.

Max and John entered the school that very night and went berserk. One or both of them scrawled messages on the walls, sprayed fire extinguishers in the hallways, and plundered the locked closet in the chemistry lab. Max carted off an assortment of chemicals and piled them into the backseat of his car.

Seth's phone rang early the next morning. It was Max; he'd left Seth a gift in his front yard. Seth walked out to find the bottles of chemicals sitting in a pile on his lawn. Panicked, he scooped them up and took them into the back, where he grabbed a shovel and started digging a hole.

His mother stepped out back and caught Seth in the act of burying the evidence.

"You know I have to tell the school now, right?" she said.

Seth was brought into the principal's office and interrogated, but he refused to name Max. One by one, the other Meridian High geeks were dragged in by the school's uniformed security officer for questioning, some in handcuffs. When it was John's turn, he spilled the beans. The school called the police, who found a telltale yellow iodine stain in the back of Max's Nissan.

The chemical theft was taken very seriously in Meridian. Max was expelled from school and prosecuted as a juvenile. He pleaded guilty to malicious injury to property, first-degree burglary, and grand theft, and spent two weeks at an in-care facility under psychiatric evaluation, where the staff diagnosed him as bipolar. His final sentence was probation. His mother sent him to Boise to live with his father and attend Bishop Kelly, the only Catholic high school in the state.

Max's first criminal conviction was a minor one. But the impulsiveness and mischievousness that spawned it ran deep in Max's personality. And he was destined to hold a lot more master keys.

2

Deadly Weapons

THIS is the Rec Room!!!!

This large, darkened room has no obvious exits. A crowd relaxes on pillows in front of a giant screen TV, and there is a fully stocked fridge and a bar.

Those words welcomed visitors to TinyMUD, an online virtual world contained in a beige computer the size of a minifridge squatting on the floor of a Pittsburgh graduate student's office. In 1990, hundreds of people from around the globe projected into the world over the Internet. Max, now a freshman at Boise State University, was one of them.

The Internet was seven years old then, and about three million people had access through a measly three hundred thousand host computers at defense contractors, military sites, and, increasingly, colleges and universities. In academia, the Net was once seen as too important to expose directly to undergraduates, but that was changing, and now any decent U.S. college allowed students online. MUDs—"multi-user dungeons"—became a favorite hangout.

Like most everything else on the pre-Web Internet, a MUD was a purely textual experience—a universe defined entirely by prose and navigated by simple commands like "north" and "south." TinyMUD was distinct as the first online world to shrug off the Dungeons and Dragons–inspired rules that had shackled earlier MUDs. Instead of limiting the power of creation to select administrators and "wizards," for example, TinyMUD granted all its inhabitants the ability to alter the world around them. Anyone could create a space of his own, define its attributes, mark its borders, and receive visitors. Inhabitants quickly anointed the user-created recreation room the world's social hub, building off it until its exits and entrances connected directly to TinyMUD spaces like Ghondahrl's Flat, Majik's Perversion Palace, and two hundred other locales.

Also gone from TinyMUD was the D & D–style reward system that emphasized collecting wealth, finishing quests, and slaying monsters. Now, instead of doing battle with orcs and building up their characters' experience points, users talked, flirted, fought, and had virtual sex. It turned out that freeing the game from the constraints of Tolkienesque roleplay made it more like real life and added to its addictive power. A common joke had it that MUD really stood for "multi-undergraduate destroyer." For Max, that would prove more than just a joke.

At Max's urging, his girlfriend Amy had joined him in one of the TinyMUDs.* The original at Carnegie Mellon University had closed in April, but by then the same free software was powering several successor MUDs scattered around the Net. Max became Lord Max, and Amy took the name Cymoril, after a tragic heroine in Michael Moorcock's Elric of Melniboné series of books and short stories—some of Max's favorites.

In the stories, Cymoril is the beloved of Elric, a weak albino transformed into a fearsome wizard emperor by dint of a magic sword called Stormbringer. To Max, the fictional sword was a metaphor for the power of a computer—properly wielded, it might turn an ordinary man into a king. But for Elric, Stormbringer was also a curse: He was bound to the sword, fought to tame it, and was ultimately mastered by it instead.

Elric's epic, doomed romance with Cymoril was very much of a piece with the fraught, uncompromising vision of romantic love Max had formed after his parents' divorce: Cymoril meets her fate during a battle between Elric and his hated cousin Yyrkoon. Cymoril pleads with

* Amy is not her real name.

Elric to sheath Stormbringer and stop the fight, but Elric, possessed by rage, presses on, striking Yyrkoon with a mortal blow. With his last breath, Yyrkoon exacts a heartbreaking revenge, pushing Cymoril onto the tip of Stormbringer.

> *Then the dark truth dawned on his clearing brain and he moaned in grief, like an animal. He had slain the girl he loved. The runesword fell from his grasp, stained by Cymoril's lifeblood, and clattered unheeded down the stairs. Sobbing now, Elric dropped beside the dead girl and lifted her in his arms.*
>
> *"Cymoril," he moaned, his whole body throbbing. "Cymoril—I have slain you."*

When she first met Max, Amy thought he was cool, rebellious, and kind of punky—different from the usual Boise crowd. But as they spent every free moment together, she began to see a darker, obsessive side to his personality, particularly after he introduced her to the Internet and TinyMUD.

At first Max was thrilled that his girlfriend shared his passion for the online world. But as Amy started making friends of her own in the MUD, including guys, he became jealous and combative. To Max it made no difference if Amy was cheating on him in the virtual world or the real one: It was cheating either way. He tried to get her to stop logging on, but she refused, and the couple began arguing online and off.

Eventually, Amy'd had enough; they were arguing about a stupid computer game? On a Wednesday night in early October 1990, the couple were in another user's room in TinyMUD when Cymoril finally told Lord Max that she wasn't sure they really belonged together after all.

It was Max's first serious relationship, and his reaction was powerful. They had sworn to spend their lives united. Now they should both die, rather than be parted, he wrote in the MUD. Then he got explicit, telling her how he'd kill her. Other users watched with growing concern as his raging took on the tone of a serious threat. What should they do?

One of the in-world wizards got Max's Internet IP address from the server—a unique identifier that was easily traced to Boise State University. The MUDers looked up the phone number for the Ada County Sheriff's Department in Boise and called in a warning that a potential murder-suicide was unfolding.

The year had begun hopefully for Max. He excelled at the part-time job his dad gave him at his computer store, HiTech Systems, performing clerical work, making deliveries in the company van, and assembling PC-compatible computers in the shop. And he managed to stay clean of probation violations—though he'd stopped taking his bipolar medication; his father didn't want him drugged, and, anyway, Max didn't agree with the diagnosis.

He began dating Amy in February of 1990, four months after meeting her at the Zoo, a dance club in Boise that catered to an underage crowd. A year younger than Max, she was blond, blue-eyed, and, when he first saw her, on the arm of Max's friend Luke Sheneman, one of the former Meridian key bearers. As Max finished up his last year in high school, they began getting serious.

Max did nothing in half measures, and his devotion to Amy was absolute. She planned on attending Boise State University, so Max applied there, postponing his dream of attending CMU or MIT. He brought her home to meet his computer, and the couple played Tetris together. Their relationship was everything his parents' hadn't been. They both thought it would never end.

His old friends barely saw him over summer break. Then the fall term began at Boise State. Max declared a major in computer science and enrolled in a battery of courses: calculus, chemistry, and a computer class on data structures. Like all students, he was given an account on the school's shared UNIX system. Like a few of them, he started hacking the computer right away. Max's path was eased by another student, David, who'd already worried his way into a bunch of the faculty accounts. They spent hours in the BSU terminal room, staring at the luminous green text of the terminals and banging on the clacky keyboards. They'd skim through faculty e-mail boxes while holding long, silent conversations, shooting messages back and forth across the room through the computer. David struggled to keep up with Max's overclocked mind and typing speed, and Max would often get impatient. "What are you waiting for?" Max would type when David fell behind in the conversation. "Respond."

A little local hacking was generally tolerated by administrators. But then Max started poking at the defenses of other Internet systems, earning him a brief ban from the BSU computer. When his access was restored, he was back on TinyMUD, fighting with Amy.

The sheriff called BSU's network administrator at two in the morning to tell him about the murder-suicide threat. The police wanted a copy of Max's computer files to examine for evidence—a request that raised difficult privacy issues for the college. After some discussion with the university's lawyer, administrators decided not to voluntarily hand over anything. Instead, they'd preserve Max's files on a computer tape and lock Max out of the computer at once.

Amy worried about what Max might do next, even as she pressed through the slow process of breaking up with him. She still cared about Max, she'd later testify, and was afraid he'd really hurt himself.

Max continued to call her after the TinyMUD incident, and the conversations followed a predictable pattern. Max would start off nice—showing the friendly, caring side that his friends and family knew well. Then he'd escalate into self-pity and threats before hanging up in anger.

On October 30, Max told Amy he wanted to talk to her in person. Still hoping to end the relationship amicably—she was bound to see Max on campus, and she didn't want him hating her—Amy agreed to come over.

Max had just moved back to his mother's home in Meridian, a ranch-style house on a quiet street a block from his old high school. He met Amy at the door, and after reassuring her that he wouldn't do anything crazy, she followed him to his bedroom at the back of his house. His mother was out, and his fourteen-year-old sister was watching TV.

His bed was still disassembled, so they sat together on the mattress on the floor and began discussing their feelings. Amy admitted that she'd met another boy in TinyMUD. His name was Chad, and he lived in North Carolina. The relationship had moved beyond the keyboard; they'd sent each other photos in the mail, and she'd been calling him on the phone.

Max struggled to control his feelings, holding back tears. He felt betrayed, he said. At the same time, he couldn't quite believe what he was hearing. He asked her for Chad's phone number, produced a calling card, and dialed his online rival.

A strained three-way conversation followed; Max introduced himself to Chad and then let Amy take over. She told Chad how she felt. Then Chad asked Amy for her phone number. She gave it to him, and the conversation drifted into an idle banter that only added to Max's agitation. He grabbed at the phone and hung it up.

Amy watched Max carefully as his breathing intensified and his eyes darted around the room.

"I'm going to kill you," he finally said. "I'm going to—you're going to die now."

She told Max that she didn't feel like she'd betrayed him, and she wouldn't apologize. Max began trembling. Then his hands were around her throat and he was pushing her down onto the mattress.

"Fine," she said. "Why don't you just kill me then?"

Once Max regained his self-control, he wanted Amy out of his sight. He pulled her from the mattress, pushed her out of his bedroom, and shuffled her through the house and out the front door.

"Go, now," he said. "Just get out, because I don't want to kill you. But I might change my mind." Amy jumped into her car and took off fast.

As she headed back to Boise, she replayed the events in her mind. Lost in thought, she didn't see the other car until she was slamming into it with a jolt and the crunch of metal against metal.

Both cars were totaled, but no one was seriously hurt. When Amy's parents learned about the confrontation at Max's house, though, they began to fear for her life. A week after the accident, Amy went to the police, and Max was arrested.

Max told his friends that Amy was exaggerating the incident. In Amy's version of events, Max had kept her prisoner in his bedroom for an hour, his hands returning to her throat repeatedly, at one point briefly cutting off her breathing. In Max's version, he'd put his fingers loosely on her throat for one minute, but he hadn't choked her, and she was always free to leave. Amy said Max continued to phone her obsessively after the incident, issuing more threats; Max said he left her alone after pushing her out of his house. As far as Max was concerned, Amy was sacrificing him to get out of trouble for her car accident.

The county prosecutor offered Max a misdemeanor deal. But a month before he was scheduled to receive a forty-five-day slap on the wrist from the judge, Max—free on his own recognizance—spotted Amy walking hand in hand with a new boyfriend down University Avenue.

Once again, Max's emotions overrode his common sense. On impulse he pulled his father's repair-shop van onto a lawn and caught up with the couple on foot. His body was tight with tension as he circled the pair.

"Hi," he said.

"You're not supposed to be around me," Amy said in protest.

"Don't you remember what we used to have?"

Amy's escort spoke up, and Max gave him a warning: "Better watch yourself, friend." Then he stalked off. A moment later, the roar of an engine. Max was back in the van, zooming across the center line toward the couple on the sidewalk. He passed close enough for Amy to feel the wind from the van as it tore off.

The deal was canceled. The district attorney stretched the law to slam Max with a felony charge of assault with a deadly weapon—his hands. It was a questionable charge: Max's hands were no deadlier a weapon than anyone else's.

The prosecution offered him a new deal: nine months in jail, if Max would admit to choking Amy. He refused. After a three-day trial, and just an hour and a half of deliberation, the jury found him guilty. On May 13, 1991, Tim Spencer and some of the other Meridian High geeks sat in the courtroom and watched as Judge Deborah Bail sentenced their friend to five years in prison.

3

The Hungry Programmers

Max found Tim Spencer's house perched at a summit in the hills separating the suburban sprawl of the San Francisco Peninsula from the quiet, undeveloped towns clinging to the Pacific coast. But "house" was too small a word. It was a villa, six thousand square feet sprawling across a fifty-acre plot overlooking the sleepy coastal town of Half Moon Bay. Max passed through the entranceway columns to the double front doors and entered the cavernous living room, where a curved wall of windows stretched from floor to ceiling.

It was a year after his parole, and Max had come to San Francisco to start over. Tim and some of his friends from Idaho had been renting the house they called "Hungry Manor," the name a reference to their first enterprise when they'd migrated to the Bay Area a year earlier. They'd planned to bootstrap into the Silicon Valley economy by forming a computer consulting business called the Hungry Programmers—will code for food. Instead, the valley quickly metabolized the geeks into full-time employment, and the Hungry Programmers morphed into an unofficial club for Tim's friends from Meridian High and the University of Idaho, two dozen in all. Hungry Manor was the group's party house and home to five of them. Max would be the sixth.

Max walked into Hungry Manor with few belongings but lots of baggage, not least a deep bitterness over his treatment by the justice system. In

1993, while Max was on his second year in prison, Idaho's Supreme Court ruled in a similar case that hands "or other body parts or appendages" couldn't be considered deadly weapons. That meant Max should never have been convicted of aggravated assault. Despite the ruling, Max's own appeal was denied on procedural grounds: The judge conceded that Max was technically not guilty of the felony for which he was serving time, but his old lawyer had failed to raise the issue in an earlier appeal, and it was too late now.

When Max was finally paroled on April 26, 1995, he left knowing that he'd served more than four years in the Idaho State Penitentiary for what, by law, should have been a misdemeanor worth sixty days in the county jail. He'd served hard time on an unjust sentence, while beyond the prison fence his friends had gone off to college, earned four-year degrees, then left Idaho to start promising careers.

He'd moved in with his dad near Seattle, and Tim, Seth, and Luke drove up from San Francisco for a reunion party of the old Meridian High geeks. They marveled at Max's prison-enhanced physique and his seemingly boundless optimism, despite having no degree and a serious felony conviction on his record. Max knew it was a time of opportunity: A British computer scientist had created the World Wide Web three months after Max's sentencing. Now there were nearly nineteen thousand websites, including one for the White House. Dial-up Internet service providers were surfacing in every major city, and America Online and CompuServe were adding Web access to their offerings.

Everyone was going online; Max was no longer the weirdo, addicted to a network nobody had heard of. Now, it turned out, he'd been at the head of a pack that was growing to include millions of people. Yet, thanks to his record, Max struggled to win computer employment in Seattle, working odd tech-support jobs through a temp agency.

Online, Max was hanging out in some rough neighborhoods. Looking for the technical challenges his day jobs denied him, Max returned to a network of chat rooms called IRC, Internet relay chat, a surviving vestige of the old Internet of his teenage years. When he'd gone to prison, IRC had been a social hotspot. But with the gentrification of the Net, most inhabitants moved uptown to easy-to-use instant messaging clients and Web-based chat systems. Those who remained on IRC tended to be either hard-core geeks or disreputable sorts—hackers and pirates scheming in the forgotten tunnels and alleyways below the whitewashed, commercialized Internet growing above them.

Max fancied himself an invisible, spectral presence in cyberspace. He chose "Ghost23" as his IRC identity—23 was his lucky number, and among other meanings it was the I Ching hexagram representing chaos. He floated into the IRC "warez" scene, where scofflaws build their reputations by pirating music, commercial software, and games. There, Max's computer skills found an appreciative audience. Max found an unprotected FTP file server at an ISP in Littleton, Colorado, and turned it into a cache for stolen software for himself and his new friends, stocked with bootlegged copies of programs like NetXray, Laplink, and Symantec's pcAnywhere.

It was a mistake. The ISP noticed the drain on its bandwidth and traced Max's uploads to the corporate offices of CompuServe in Bellevue, where Max had just started working a new temp job. Max was fired. Barely a year after his release from prison, his name was mud.

That was when Max decided to start over again in Silicon Valley, where the dot-com economy was swelling to ripeness and a talented computer genius could pick up work without a lot of questions about his past.

He'd need a new name, unstained by his past folly. Max had been known by a nickname in the joint, one abbreviated from a cyberpunk-themed 'zine he'd published from the prison typewriter: *Maximum Vision*. It was a clean, optimistic name that exemplified everything he wanted to be and crystallized his clarity and hopefulness.

As he left Seattle in the rearview mirror, he said good-bye to Max Butler. From now on, he would be Max Ray Vision.

. . .

Max Vision found that life in Hungry Manor was good. Surrounded by rolling meadows on all sides, the house boasted two wings, four bedrooms, a maid's quarters, a full dining room, a livestock pen, and a brick pizza oven and indoor barbecue in a vented room adjoining the vast, sunlit kitchen. The Hungries had turned the library into a computer lab and server room, packing in a slew of custom-built gaming PCs for recreation. They ran networking cable into every room and energized it with a high-speed Internet link that necessitated the partial shutdown of the 92 freeway as the phone company trenched a new cable run alongside the road. A vintage phone system linked the west wing to the east. As a finishing touch, one of the Hungry Programmers had brought in a hot tub and set it up on the grounds, under the stars.

Max couldn't have asked for a better launchpad for his new life. One of the resident Hungries got him a job as a system administrator at MPath Interactive, a computer gaming start-up in Silicon Valley that was flush with venture capital. He threw himself into the job. Defying the stereotype of a computer nerd, he drew his greatest satisfaction from his support duties. He liked helping people.

But it wasn't long before Max's antics in Seattle caught up with him. One morning, a process server showed up at his cubicle to hand him a $300,000 lawsuit filed by the Software Publishers Association—an industry group that had decided to use his piracy bust to send a message. "This action is a warning to Internet users who believe they can infringe software copyrights without fear of exposure or penalty," the association proclaimed in a press release.

As the first lawsuit of its kind, the case earned Max Butler a brief write-up in *Wired* magazine and a mention in a congressional hearing on Internet piracy. Max Vision, though, emerged largely unscathed—few in his new life made the connection to the man named in the high-profile lawsuit.

When the press attention faded, the SPA was willing to quietly settle the case for $3,500 and some free computer consulting. The whole affair even had a silver lining. It introduced Max to the FBI.

Chris Beeson, a young agent with the bureau's San Francisco computer crime squad, gave Max his pitch. The FBI could use Max's assistance navigating the computer underground. Recreational hackers were no longer a target for the bureau, he said. There was a new, more dangerous breed of computer criminal emerging: "real" criminals. They were cyberthieves, pedophiles, even terrorists. The FBI was no longer chasing people like Max and his ilk. "We're not the enemy," said Beeson.

Max wanted to help, and in March 1997 he was formally inducted into the FBI's Criminal Informant program. His first written report for the bureau was an introductory course on the virus-writing, warez, and computer-hacking scenes. His follow-up report ten days later ran down compromised file-transfer sites—like the one he'd exploited in Seattle—and a music piracy gang called Rabid Neurosis that had debuted the previous October with a bootlegged release of Metallica's *Ride the Lightning.*

When Max got his hands on a pirated version of AutoCAD that was being circulated by a crew called SWAT, the FBI rewarded him with a $200 payment. Beeson had Max sign the receipt with the bureau's code name for its new asset: Equalizer.

Max liked the FBI agent, and the feeling seemed to be mutual. Neither of them knew that Chris Beeson would one day put his Equalizer back behind bars and begin Max's transformation into one of the "real" criminals Beeson had hoped to catch.

4

The White Hat

Max was building his new life at a time of profound change in the hacking world.

The first people to identify themselves as hackers were software and electronics students at MIT in the 1960s. They were smart kids who took an irreverent, antiauthoritarian approach to the technology they would wind up pioneering—a scruffy counterweight to the joyless suit and lab-jacket culture then epitomized by the likes of IBM. Pranks were a part of the hacker culture, and so was phone phreaking—the usually illegal exploration of the forbidden back roads of the telephone network. But hacking was above all a creative effort, one that would lead to countless watershed moments in computer history.

The word "hacker" took on darker connotations in the early 1980s, when the first home computers—the Commodore 64s, the TRS-80s, the Apples—came to teenagers' bedrooms in suburbs and cities around the United States. The machines themselves were a product of hacker culture; the Apple II, and with it the entire home computer concept, was born of two Berkeley phone phreaks named Steve Wozniak and Steve Jobs. But not all teenagers were content with the machines, and in the impatience of youth, they weren't inclined to wait for grad school to dip into real processing power or to explore the global networks that could be reached with a phone call and the squeal of a modem. So they began illicit forays into corporate, government, and academic systems and took their first tentative steps into the ARPANET, the Internet's forerunner.

When those first young intruders began getting busted in 1983, the national press cast about for a word to describe them and settled on the one the kids had given themselves: "hackers." Like the previous generation of hackers, they were pushing the limits of technology, outwitting the establishment, and doing things that were supposed to be impossible. But for them, that involved breaching corporate computers, taking over telephone switches, and slipping into government systems, universities, and defense contractor networks. The older generation winced at the comparison, but from that point on, the word "hacker" would have two meanings: a talented programmer who pulled himself up by his own bootstraps, and a recreational computer intruder. Adding to the confusion, many hackers were both.

Now, in the mid-1990s, the hacking community was dividing again. The FBI and the Secret Service had staged arrests of high-profile intruders like Kevin Mitnick and Mark "Phiber Optik" Abene, a New York phone phreak, and the prospect of prison stigmatized recreational intrusion while raising the risk far beyond the rewards of ego and adventure. The impetus for cracking computers was fading as well: The Internet was open to anyone now, and personal computers had grown powerful enough to run the same operating systems and programming languages that fueled the big machines denied to amateurs. Most of all, there was real money to be made defending computers and none attacking them.

Cracking systems was becoming uncool. Those possessed of a hacker's mind-set were increasingly rejecting intrusion and going right into legitimate security work. And the intruders started hanging up their black hats to join them. They became the "white-hat hackers"—referencing the square-jawed heroes in old cowboy films—applying their computer skills on the side of truth and justice.

Max thought of himself as one of the white hats. Watching for new types of attacks and emerging vulnerabilities was now in his job description, and as Max Vision, he was beginning to contribute to some of the computer-security mailing lists where the latest developments were discussed. But he couldn't completely exorcise Ghost23 from his personality. It was an open secret among Max's friends that he was still cracking systems. When he saw something novel or interesting, he saw no harm in trying it out for himself.

Tim was at work one day when he got a call from a flummoxed system administrator at another company who'd traced an intrusion back to Hungry.com—the online home of the Hungry Programmers, where they hosted their projects, hung their résumés, and maintained e-mail addresses that would remain steady through job changes and other upheavals. There were dozens of geeks on the shared system, but Tim knew at once who was responsible. He put the sysadmin on hold and phoned up Max.

"Stop. Hacking. Now," he said.

Max stammered out an apology—it was the burning lawn all over again. Tim switched back to the other line, where the system administrator happily reported that the attack had stopped in its tracks.

The complaint surprised and confused Max—if his targets knew what a good guy he was, they wouldn't take issue with some harmless intrusions. "Max, you gotta get permission," Tim explained. He offered some life advice. "Look, just sort of imagine that everyone's looking at you. That's a good way to ensure that what you're doing is correct. If I was standing there, or your dad was standing there, would you still feel the same about doing it? What would we say?"

If there was one thing Max was missing in his new life, it was a partner to share it with. He met twenty-year-old Kimi Winters at a rave called Warmth, held on an empty warehouse floor in the city—Max had become a fixture in the rave scene, dancing with a surprising, fluid grace, whirling his arms like a Brazilian flame dancer. Kimi was a community college student and part-time barista. A foot shorter than Max, she sported an androgynous appearance in the shapeless black hoodie she liked to wear when she went out. But on a second look, she was decidedly cute, with apple cheeks and her Korean mother's copper-tinted skin. Max invited Kimi to a party at his place.

The parties at Hungry Manor were legendary, and when Kimi arrived the living room was already packed with dozens of party guests from Silicon Valley's keyboard class—programmers, system administrators, and Web designers—mingling under the glass chandelier. Max lit up when he spotted her. He led her on a tour of the house, pointing out the geeky accoutrements the Hungry Programmers had added.

The tour ended in Max's bedroom in Hungry Manor's east wing. For all of the grandeur of the house, Max's room had the charm of a monk's cell—no furniture but a futon on the floor, no comforts except a computer. For the party, Max had trained blue and red spotlights on a bottle of peppermint schnapps—his only vice. Kimi returned for dinner the next night, and there was a single item on his vegetarian menu: raw cookie dough. Max shaved the sugary sludge off in slices and served it to his date with the schnapps. Why, after all, would anyone *not* eat raw cookie dough for dinner, given the option?

Kimi was intrigued. Max needed so little to be happy. He was like a child. When his birthday came soon after the party, she sent a decorated box of balloons to his office at MPath, and Max was moved nearly to tears by the gesture.

She was his "dream girl," he told her later. They began to talk about committing to a life together.

In September, Hungry Manor's landlord, unhappy with the programmers' upkeep of the estate, reclaimed the house, and after a final bash to bid farewell to their communal mansion, the Hungries scattered to rentals throughout the Bay Area. Max and Kimi landed in their own place in Mountain View, a cramped studio in a barracks-like apartment complex alongside the 101 freeway, Silicon Valley's congested main artery.

Max resumed his work for the FBI, and his haunting of IRC led him to a new opportunity—his chance to break out as a white-hat hacker. He'd made a friend in the chat rooms who was starting a real consulting business in San Francisco and was interested in bringing Max on board. Max went up to the city to visit Matt Harrigan, aka, "Digital Jesus."

Harrigan, just twenty-two, was one of four white hats who'd been profiled in a *Forbes* cover story the previous year, and he'd cannily used his fifteen minutes of fame to win some seed money for a business: a professional hacking shop in San Francisco's financial district.

The idea was simple: Corporations would pay his company, Microcosm Computer Resources, to put their networks through a real hack attack, culminating in a detailed report on the client's security strengths and weaknesses. The business of "penetration testing"—as it was called—had been dominated by the Big Five accounting firms, but Harrigan was already signing up clients by admitting something that no accounting firm would ever announce: that his experience came from real-life hacking, and he was freely hiring other ex-hackers.

MCR would be billing out between $300 and $400 an hour, Harrigan explained. Max would work as a subcontractor, making $100 to $150. All for doing two of the things he liked most in the world: hacking into shit and writing reports.

Max had found his niche. It turned out his single-mindedness made him a natural at penetration testing: He was immune to frustration, hammering at a client's network for hours, moving from one attack vector to another until he found a way in.

With Max making real money at MCR, Kimi quit her job as a barista and found more rewarding work teaching autistic students. The couple moved from the cramped apartment in Mountain View to a duplex in San Jose. In March, they got married in a church on a college campus in Lakewood, Washington, where Kimi's family lived.

Tim Spencer and most of the Hungry Programmers went up to Washington to see their problem child married off. Max's parents, his sister, Kimi's family, and scores of friends and extended family showed up for the ceremony. Max wore a tuxedo and a broad grin, and Kimi glowed in her white wedding dress and veil. Surrounded by family and beloved friends, they were a picture-perfect young couple beginning a life together.

They posed outside: Kimi's father, a military man, stood proudly in his dress uniform, her mother in a traditional Korean *hanbok*. Flanked by his own parents, Max beamed at the camera, while storm clouds gathered overhead in the Pacific Northwest sky.

It was three years almost to the day since Max walked out of prison, and he had everything now—a devoted wife, a promising career as a white-hat hacker, a nice home. In just a few weeks, he'd throw it all away.

5

Cyberwar!

Back home in San Francisco, a temptation was waiting for Max, written in computer code.

```
bcopy(fname, anbuf, alen = (char *)*cpp - fname);
```

It was one line of nine thousand comprising the Berkeley Internet Name Domain, an ancient girder in the Internet's infrastructure, as important as any router or fiber-optic cable. Developed in the early 1980s with a grant from the Pentagon's Defense Advanced Research Projects Agency (DARPA), BIND implemented the scalable Domain Name System, a kind of distributed telephone directory that translates strings like Yahoo.com, which humans understand, into the numeric addresses the network comprehends. Without BIND, or one of the competing programs that followed, we'd be getting our online news from 157.166.226.25 instead of CNN.com and visiting 74.125.67.100 to perform a Google search.

BIND was one of the innovations that made the explosive growth of the Internet possible—it replaced a crude mechanism that couldn't have expanded with the Net. But in the 1990s, it was also one of the legacy programs that were shaping up as the modern Internet's biggest security problem. The code was a product of a simpler time, when the network was cloistered and threats were few. Now hackers were plumbing its depths and coming back with a seemingly endless supply of security holes.

A high priesthood of network experts called the Internet Software Consortium appointed themselves keepers of the code and had begun furiously rewriting it. But in the meantime, the most modern, sophisticated networks in the world, with sparkling new servers and workstations, were running a buggy computer program from another age.

In 1998, security experts discovered the latest flaw in the code. It boiled down to that single line. It accepted an inquiry from the Internet, as it should, and copied it byte for byte into the temporary buffer "anbuf" in the server's memory. But it didn't properly check the size of the incoming data. Consequently, a hacker could transmit a deliberately overlong query to a BIND server, overflow the buffer, and spill data into the rest of the computer's memory like oil from the *Exxon Valdez*.

Performed haphazardly, such an attack would cause the program to crash. But a careful hacker could do much worse. He could load the buffer with his own small snippet of executable computer code, then he could keep going, tripping cautiously all the way to the top of the program's memory space, where a special short-term storage area called the "stack" resides.

The stack is where the computer's processor keeps track of what it's doing—every time a program diverts the computer off to a subroutine, the processor pushes its current memory address onto the stack, like a bookmark, so it knows where to return to when it's done.

Once a hacker is in the stack, he can overwrite the last return address with the location of his own malicious payload. When the computer is done with the current subroutine, it returns not to where it began, but to the hacker's instruction—and because BIND runs under the all-powerful administrative "root" account, the attacker's code does as well. The computer is now under the hacker's control.

Two weeks after Max and Kimi's wedding, the government-funded Computer Emergency Response Team at Carnegie Mellon University—which runs a kind of Emergency Broadcast System for security holes—

issued an alert about the BIND flaw, along with a link to the simple fix: two additional lines of computer code that rejected overlong queries. But CERT packaged its alert with two other BIND vulnerabilities that were of little consequence and understated the importance of the hole. Consequently, not everyone appreciated the gravity of the situation.

Max understood perfectly.

He read the CERT advisory with amazement. BIND came installed standard with Linux, and it ran on servers on corporate, ISP, nonprofit, educational, and military networks. It was everywhere. And so was the defective line of code. The only thing holding back a feeding frenzy of attacks was that nobody had written a program to exploit the security hole. But that was just a matter of time.

Sure enough, on May 18, an exploit program showed up on Rootshell .com, a computer security news site run by hobbyists. Max picked up the phone and called his FBI contact, Chris Beeson, at home. The situation was serious, he explained. Anybody who hadn't installed the BIND patch could now be hacked by any script kiddy capable of downloading a program and typing a command.

If history was a guide, government computers would be particularly vulnerable. Just a month earlier, a less serious bug in the Sun Solaris operating system had led to a hacker cracking computers at a dozen U.S. military bases, in what a deputy defense secretary called "the most organized and systematic attack to date" on American defense systems. Those attacks had set off a full-blown cyberwarfare false alarm: The Pentagon gave the intrusions the code name "Solar Sunrise" and considered Saddam Hussein the prime suspect until investigators traced the attacks to a young Israeli hacker who was just playing around.

Max called Beeson again the next day, when a hacker group named ADM released a weaponized version of the BIND exploit designed to scan the Internet at random looking for unpatched servers, then break in, install itself, and use the newly compromised computer as a platform for still more scans and break-ins. It was a certainty now that someone was going to own the entire Internet. It was just a question of who.

He hung up and pondered. *Someone* was going to do it. . . .

He shared his plans with his new wife in boyish, excited tones. Max would author his own BIND attack. His version would close the hole everywhere it found it, like releasing sterile fruit flies to tamp down an infestation. He would limit his attack to the targets most in need of an emergency security upgrade: U.S. military and civilian government sites.

"Don't get caught," said Kimi, who'd learned not to argue with Max when he was like this, his mind hostage to an idea.

Max was struggling with the binary nature of his personality: the professional married man with a stake in the world around him, and the impulsive child tempted by every call to mischief. The child won. He sat at his keyboard and plunged into furious programming.

His code would operate in three rapid-fire stages. It would begin by flinging a virtual grappling hook through the BIND hole, executing commands that forced the machine to reach out over the Internet and import a 230-byte script. That script, in turn, would connect it to a different host infiltrated by Max, where it would download a hefty package of evil called a "rootkit."

A rootkit is a bundle of standard system programs that have been corrupted to secretly serve the hacker: A new login program operates just like the real thing but now includes a back door through which the intruder can reenter the machine. The "passwd" program still lets users change their passwords but also quietly records and stores the new password where it can be retrieved later. The new list program lists the contents of a directory, as it should, but takes care to conceal any files that are part of the rootkit.

Once the rootkit was in place, Max's code would accomplish what the government failed to do: It would upgrade the hacked computer to the latest version of BIND, closing the security hole through which it had entered. The computer would now be safe from any future attacks, but Max, the benevolent meddler, would still be able to reenter the system at will. Max was at once fixing the problem and exploiting it; he was a black hat and a white hat at the same time.

The whole attack would take just a couple of minutes each time. One moment, the computer would be controlled by the system administrators; then, grappling hook, download script, rootkit, and it was in Max's pocket.

Max was still programming when the FBI got back to him and asked for a full report on the BIND hole. But the feds had had their chance; Max's code would speak for him now. He took a moment to crack a couple of college machines to use as a staging ground, then, on May 21, a Tuesday, he dialed the Internet through a stolen Verio account . . . and launched.

The results were instant and highly satisfying. Max's grappling-hook code was designed to signal its success to his computer over the Verio dial-up, so he could watch the attack spread. Hacked machines around the country reported back to him, an Xterm window popping up on his screen for each one. Brooks Air Force Base—now property of Max Vision. McChord, Tinker, Offutt, Scott, Maxwell, Kirtland, Keesler, Robins. His code wormed into Air Force servers, Army computers, a machine in the office of a cabinet secretary. Each machine now had a back door that Max could use any time he wanted.

Max was notching up military conquests like points in a video game. When his code swept into the Navy's Internet space, it found so many unpatched BIND servers that the stream of pop-ups turned into a torrent. His own computer struggled under the strain, then crashed.

After some fine-tuning, he relaunched. For five days he was absorbed in his growing dominion over cyberspace. He ignored e-mail from the

FBI, who still wanted that report. "Where's the stuff?" Agent Beeson wrote. "Please call."

There had to be more he could do with the power to crack almost any network he wanted. Max trained his BIND exploit on the servers of Id Software in Mesquite, Texas, a gaming company developing a third installment of the enormously popular first-person shooter Quake. Max loved first-person shooters. He was on the network in a flash, and after some exploring, he emerged with his trophy. He announced to Kimi that he'd just obtained the source code—the virtual blueprints—for Quake III, the most anticipated game of the year.

Kimi was unmoved. "Can you put it back?"

Max soon realized that his attacks were getting some attention. At Lawrence Berkeley National Laboratory, a researcher named Vern Paxson spotted Max's scanning using a new system he'd developed called BRO, for Big Brother. BRO was an experiment in a relatively new kind of security countermeasure called an intrusion detection system—a cyber burglar alarm with the sole function of sitting quietly on a network and sifting through all the traffic for suspicious activity, alerting administrators when it spots something that doesn't look right.

Paxson wrote a full report on the attack for CERT. Max intercepted it and was impressed. The researcher had not only detected his attack, he'd compiled a list of servers that Max's code was attacking through Lawrence Berkeley's network—Max was using the network as one of his secondary launch points. He sent Paxson an anonymous note from the lab's root account.

```
Vern,

I'm sorry to have caused you any inconvenience,
but I single-handed fixed a MAJOR GAPING SECURITY
HOLE in many of your systems. I admit there were
new holes but these were all passworded, and I
```

would never cause damage to someone's computer system.

If I didn't hit these, someone else would have, and they would have been dirty. These kids leave warez and IRC BS laying everywhere, and /bin/rm systems when they are unhappy. Lame.

You might not appreciate what I was doing, but it was for the greater good. I am abandoning all hosts on that list that you captured. . . . I am not touching those systems since I know you turned them over to CERT. CERT should hire people with my skill. Of course, if paid I would never leave rootkits or such.

Pretty clever though? Heh. It was a blast. Owning hundreds, nay thousands of systems, and knowing that you were FIXING them on the way . . .

Uhm, I'm not ever doing this sort of shit again. You have my tools now. That pisses me off . . .

Hrm. Anyway I just don't want this to happen again, so I'm going to let it lie . . .

"The Cracker"

With that, Max shut down his five-day attack on the government, with more cracked systems behind him than he could count. He was satisfied that he'd made the Internet safer than it was before; thousands of computers that had been vulnerable to every hacker in the world were now vulnerable to only one: Max Vision.

Max immediately jumped into a new, more socially acceptable project: He would write a Web application that would let anyone on the Internet request an automatic real-time scan of their network to assess whether or not they were open to the BIND attack. He also conceived a benign variant of the siege he'd just concluded. Like before, he would scan government and military networks. But instead of cracking the vulnerable computers, he'd automatically send an e-mail warning to the administrators. There'd be no need to hide behind a hacked dial-up account this time. Both services would live on his brand-new public website: Whitehats.com.

After two days and nights of work, he was knee-deep in his new, legal hacking project when Beeson e-mailed again. "What happened? Thought you'd send me e-mail."

Max could hardly explain to his FBI friend that he'd been busy staging one of the largest government computer breaches in history. So he emphasized his new project instead. "I am almost finished creating a public service vulnerability scanner and patch site—but there are some parts that aren't ready for release," he wrote back.

"Oh, and here is the ADM worm program," he added. "I don't think it will spread very far."

6

I Miss Crime

On the afternoon of June 2, Max opened the door of his San Jose duplex to greet Chris Beeson and registered instantly that he was in trouble: There were three other suits with the FBI agent, including Beeson's surly boss, Pete Trahon, head of the computer crime squad.

The month after the BIND attack had been a busy time for Max. He launched Whitehats.com, and it was an instant success in the security world. In addition to housing his scanning tool, the site collected the latest CERT advisories and links to BIND software patches, as well as a paper Max had written dissecting the ADM worm with the clarity and the discerning eye of a connoisseur. Nobody in the community suspected that Max Vision, the rising star behind Whitehats.com, had personally provided the brightest example of the seriousness of the BIND security hole.

He was also continuing to file reports to the FBI. After his last one, Beeson began e-mailing to arrange a casual meeting, supposedly to go over Max's latest findings. "How 'bout if we just meet at your place?" Beeson wrote. "I know I have the address somewhere around here."

Now that he was on Max's doorstep, Beeson explained why they were really there. He knew all about Max's attack on the Pentagon. One of the men with him, a young Washington, DC–based Air Force investigator named Eric Smith, had traced the BIND intrusions to Max's house. Beeson had a search warrant.

Max let them in, already apologizing. He only meant to help, he explained.

They chatted amicably. Max, happy for an audience, grew expansive, describing the twists and turns of his attack and listening with interest as Smith described how he'd tracked Max through the pop-up messages Max had used to alert himself when a system was subverted: The messages went to a Verio dial-up, and a subpoena to the ISP produced Max's phone number. It hadn't been difficult. Max had convinced himself he was doing something positive for the Internet, so he hadn't done much to cover his tracks.

The feds asked if anyone had known what Max was up to, and he said his boss was involved. Matt Harrigan—Digital Jesus—had not completely given up hacking himself, Max said, adding that Harrigan's company was about to get a contract with the National Security Agency.*

At the agents' behest, Max wrote out a confession. "My motives were purely for research and 'to see if it could be done,'" Max wrote. "I know this is no excuse, and believe me, I am sorry for it, but it's the truth."

Kimi came home from school to find the feds still tossing the house. Like grazing deer, they looked up in unison as she entered, dismissed her as unthreatening, and turned wordlessly back to their work. When they left, they hauled Max's computer equipment with them.

The door closed, leaving the newlyweds alone in what was left of their home. An apology formed on Max's lips. Kimi cut him off angrily.

"I told you not to get caught!"

The FBI agents saw an opportunity in Max's crime. Trahon and Beeson returned to Max's home and gave their former ally the score. If Max

---

* Harrigan's involvement is in dispute. Max says he planned the BIND attack with Harrigan at the MCR office and that Harrigan wrote the program that built the target list of government computers. Harrigan says he was not involved but was aware of what Max was up to.

hoped for leniency, he'd have to work for them—and writing reports wasn't going to cut it anymore.

Eager to make amends and determined to salvage his life and career, Max didn't ask for anything in writing. He took it on faith that if he helped the FBI agents, they would help him.

Two weeks later, Max got his first assignment. A gang of phone phreaks had just hijacked the phone system at the networking company 3Com and were using it as their own private teleconferencing facility. Beeson and Trahon could dial into the illicit chat line, but they doubted their ability to blend in with the hackers and gain any useful intelligence. Max studied up on the latest phone phreaking methods, then dialed into the system from the FBI's field office while the bureau recorded the call.

Dropping the names of hackers he knew and drawing on his own expertise, Max easily persuaded the phone phreaks that he was one of them. They opened up and revealed that they were an international gang of about thirty-five phone hackers called DarkCYDE, living mostly in Britain and Ireland. DarkCYDE aspired to "unite Phreakers and Hackers all over the world into one big digital army," according to the group's blustery manifesto. But at root they were just kids playing with the phone, just as Max had done in high school. After the call, Beeson asked Max to stay close to the gang. Max chatted them up on IRC and turned over the logs to his handlers.

Pleased with Max's work, the agents summoned him to the federal building in San Francisco a week later to brief him on a new assignment. This time, he'd be going to Vegas.

Max's eyes moved over the nest of linen-clad card tables in the gaudy exhibit hall of the Plaza Hotel and Casino. Dozens of young men in T-shirts and shorts or jeans—the hacker's uniform—were at the tables hunkered over a bank of computer workstations or standing on the sidelines, occasionally pointing at something on a screen.

To the untrained eye, it was a strange way to spend a weekend in Sin City—banging on keyboards like some anonymous cubical drone, far from the pool, the slots, and the shows. But the hackers were in pitched competition, working in teams to penetrate a clutch of computers hanging off a hastily erected network. The first team to leave their virtual marker in one of the targets would claim a $250 prize and valuable bragging rights—with points also awarded for hacking other competitors. New attacks and ruses were flowing from the hackers' fingers, and secret, stockpiled exploits were being pulled from virtual armories to be used in public for the first time.

At Def Con, the world's largest hacking convention, the Capture the Flag competition was Fischer vs. Spassky every year.

Kimi wasn't impressed, but Max was in heaven. Across the floor, more tables were cluttered with vintage computer gear, odd electronics, lock-picking tools, T-shirts, books, and copies of *2600: The Hacker Quarterly.* Max spotted Elias Levy, a famous white-hat hacker, and pointed him out to Kimi. Levy, aka Aleph One, was the moderator of the Bugtraq mailing list—the *New York Times* of computer security—and the author of a seminal tutorial on buffer overflows called "Smashing the Stack for Fun and Profit" that had appeared in *Phrack.* Max didn't dare approach the luminary. What would he say?

Max wasn't the only law enforcement mole at Def Con, of course. From its humble beginnings in 1992 as a one-off conference pulled together by a former phone phreak, Def Con had grown into a legendary gathering that drew nearly two thousand hackers, computer security professionals, and hangers-on from around the world. They came to party in person with comrades they'd befriended online, present and attend technical talks, buy and sell merchandise, and get very, very drunk in all-night bashes in the hotel rooms.

Def Con was such an obviously target-rich environment for the government that the organizer, Jeff "the Dark Tangent" Moss, had invented a new convention game called Spot the Fed. A hacker who thought he'd identified a G-man in the crowd could point him out, make a case, and, if the audience concurred, take home a coveted I SPOTTED THE FED AT DEF CON T-shirt. Often the suspected fed would just give up and good-naturedly whip out a badge, giving the hacker an easy win.

Max's mission was broad. Trahon and Beeson wanted him to chum up to his fellow hackers and try to get their real names, then lure them into exchanging public PGP encryption keys, which security-minded geeks use like sealing wax to encrypt and sign their e-mail. Max's heart just wasn't in it. Writing reports for the bureau was one thing, and he'd had no qualms about getting the goods on the DarkCYDE phreaks, who were too young to get in real trouble. But this assignment smelled like snitching. Personal loyalty was written deep into Max's firmware, and one look at the Def Con crowd told him these were his people.

Many of the hackers were reluctantly giving up childish things, migrating into legitimate dot-com jobs or starting security companies. They were becoming white hats, like Max. A popular T-shirt at the conference summed up the mood: I MISS CRIME.

Max shrugged off the FBI's edict and began attending the parties and the talks. On the roster this year was a much-anticipated software release by the Cult of the Dead Cow. The cDc were the rock stars of the hacker world—literally: They recorded and performed music and infused their conference presentations with over-the-top theatrics that made them media darlings. At this Def Con the group was unleashing Back Orifice, a sophisticated remote-control program for Windows machines. If you could trick someone into running Back Orifice, you could access their files, see what was on their screen, and even look through their webcam. It was designed to embarrass Microsoft for the shoddy security in Windows 98.

The crowd at the Back Orifice presentation was ecstatic, and Max found the energy infectious. But of more pragmatic interest to Max was a talk on the legalities of computer hacking by a San Francisco criminal defense attorney named Jennifer Granick. Granick opened her presentation by describing the recent landmark prosecution of a Bay Area hacker named Carlos Salgado Jr., a thirty-six-year-old computer repairman who, more than any other hacker, represented the future of computer crime.

From his room in his parents' house in Daly City, a few miles south of San Francisco, Salgado had cracked a major technology company and stolen a database of eighty thousand credit card numbers, with names, ZIP codes, and expiration dates. Credit card numbers had been hacked before, but what Salgado did next assured him a place in the cybercrime history books. Using the handle "Smak," he jumped into the #carding chat room on IRC and put the entire list up for sale.

It was like offering a 747 for sale at a flea market. At the time, the online credit card fraud underground was a depressing bog of kids and small timers who'd barely advanced beyond the previous generation of fraudsters fishing receipt carbons from the Dumpsters behind the mall. Their typical deals were in the single digits, and their advice to one another was tainted by myth and idiocy. Much of the conversation unfolded in an open channel where anyone in law enforcement could log in and watch—the carders' only security was the fact that nobody would bother.

Remarkably, Salgado found a prospective buyer in #carding—a San Diego computer science student who'd been putting himself through college by counterfeiting credit cards, getting the account numbers from billing statements pilfered from the U.S. mail. The student had mob contacts who, he believed, would buy Smak's entire stolen database for six figures.

The deal went south when Salgado, looking to perform a little due diligence, hacked his customer's ISP and poked through his files. When the student found out, he got mad and secretly began working with the FBI. On the morning of May 21, 1997, Salgado showed up at a meeting with his buyer at the smoking lounge at San Francisco International Airport, where he expected to trade a CD-ROM containing the database for a suitcase packed with $260,000 in cash. Instead, he was arrested by the San Francisco computer crime squad.

The foiled plot was an eye-opener for the FBI: Salgado represented the first of a new breed of profit-oriented hacker, and he posed a threat to the future of e-commerce. Surveys showed that Web users were anxious about sending credit card numbers into the electronic ether—it was the number one thing holding them back from Internet purchasing. Now, after years of struggling to gain consumers' trust and reward the faith of investors, e-commerce companies were starting to win over Wall Street. Less than two weeks before Salgado's arrest, Amazon.com had launched its long-awaited initial public offering and ended the day $54 million richer.

Salgado's IPO was higher: The credit card companies determined the total spending limits on his eighty thousand cards amounted to over a billion dollars—$931,568,535 if you subtracted the legitimate owners' outstanding balances. The only thing he'd been missing was a NASDAQ to trade on. Once the underground figured out that part of the equation, it would be an industry of its own.

As soon as Salgado was arrested, he'd confessed everything to the FBI. That, Granick told the Def Con hackers in her presentation, was his big mistake. Despite his cooperation, Salgado had been sentenced to thirty months in prison earlier that year.

"Now, the FBI wanted me to tell you that it was good for Mr. Salgado that he talked." Granick paused. "That's bullshit.

"Just say no!" she said, and cheers and whistles swelled from the audience. "There's never any good reason to talk to a cop. . . . If you're going to cooperate, you're going to cooperate after consulting with a lawyer and cutting a deal. There's never any reason to give them information for free."

In the back of the room, Kimi prodded Max in the ribs with her elbow. Everything Granick was advising computer intruders not to do, Max had done. Everything.

Max was having second thoughts about his arrangement with the feds.

. . .

"We need to make some changes in the way we do business."

Max could feel the frustration radiating from his screen as he read the latest note from Chris Beeson. Max had returned from Def Con empty-handed and then blown off a meeting at the federal building at which he was supposed to get a new assignment, pissing off Beeson's supervisor, Pete Trahon. Continuing his e-mail, Beeson warned Max of dark consequences for continued flakiness. "In the future, missed appointments without exceptional reasons will be considered uncooperative on your part. If you are not willing to cooperate then we HAVE to take the appropriate actions. Pete is meeting with the prosecutor on YOUR case Monday. He wants to meet with you promptly in our office at 10:00am sharp, MONDAY 8/17/98. I am not available next week (that is why I wanted to meet with you this week) so you're going to have to deal directly with Pete."

This time, Max showed up. Trahon explained that he'd become interested in Max's boss at MCR, Matt Harrigan. The agent was alarmed at the idea of a hacker running a cybersecurity shop staffed with other hackers, like Max, and vying for a contract with the NSA. If Max wanted to make the FBI happy, he had to get Harrigan to admit he was still hacking and had played a role in Max's BIND attack.

The agent gave Max a new form to sign. It was Max's written consent to wire him for sound. Trahon handed him a bureau-issued recording device disguised as a pager.

On the way home, Max pondered the situation. Harrigan was a friend and fellow hacker. Now the FBI was asking Max to perform the ultimate betrayal—to become Digital Jesus's real-life Judas.

The next day, Max met Harrigan at a Denny's diner in San Jose, without the FBI wire. His eyes scanned over the other diners and looked out the window into the parking lot. There could be feds anywhere.

He pulled out a piece of paper and slipped it across the booth. "Here's what's going on. . . ."

Max phoned Jennifer Granick after the meeting—he'd gotten her card at the conclusion of her Def Con talk—and she agreed to represent him.

When they learned Max had lawyered up, Beeson and Trahon wasted no time in officially dropping him as an informant. Granick began phoning the FBI and the prosecutor's office to find out what the government had planned for her new client. Three months later she finally got an answer from the government's top cybercrime prosecutor in Silicon Valley. The United States was no longer interested in Max's cooperation. He could look forward to going back to prison.

7

Max Vision

With his government service at an end, Max went to work building his reputation as a white-hat hacker, even as he lived under the sword of Damocles of a pending federal indictment.

The BIND vulnerability and the resultant success of Whitehats.com had given him a running start. Now Max hung up his own shingle as a computer security consultant, erecting a new website touting his services as a hacker for hire at one hundred dollars an hour—or free to nonprofit groups. His chief selling point: a 100 percent success rate in penetration tests. He had never once confronted a network he couldn't crack.

It was an exciting time to be a white hat. The rebellious spirit that drove the open-source software movement was planting itself in the computer security world, and a new crop of college graduates, dropouts, and former and current black hats was upending the conservative assumptions that had dominated security thinking for decades.

First to be dustbinned was the tenet that security holes and attack methods should be kept quiet, held privately among a cadre of trusted responsible adults. The white hats called this notion "security through obscurity." The new generation preferred "full disclosure." Discussing security problems widely not only helped get them fixed, but it also advanced the science of security, and hacking, as a whole. Keeping bugs private only benefited two groups: the bad guys who were exploiting them, and vendors like Microsoft that preferred to fix security holes without confessing the details of their screwups.

The full-disclosure movement spawned the Bugtraq mailing list, where hackers of any hat color were encouraged to send in detailed reports of security flaws they'd found in software. If they could provide an "exploit"—code that demonstrated the flaw—so much the better. The preferred path to full disclosure was to first notify the software maker and give that company time to issue a patch before releasing the flaw or exploit on Bugtraq. But Bugtraq didn't censor, and it was common for a bug finder to drop a previously unknown exploit onto the list, releasing it simultaneously to thousands of security researchers and hackers in the span of minutes. The maneuver was all but guaranteed to kick a software company into rapid response.

Bugtraq provided hackers with a way to show off their expertise without breaking the law. The ones who were still cracking systems had an invigorated white-hat community to deal with, armed with a growing arsenal of defensive tools.

In late 1998, a former NSA cybersecurity contractor named Marty Roesch developed one of the best. Roesch thought it would be fun to see what random attacks were crossing his home cable modem connection while he was at work. As a weekend project, he cranked out a packet sniffer called Snort and released it as an open-source project.

At first, Snort was nothing special—a packet sniffer is a common security tool that eavesdrops on the traffic crossing a network and dumps it to a file for analysis. But a month later, Roesch turned his program into a full-blown intrusion detection system (IDS), which would alert the operator whenever it spotted network traffic that matched the signature of a known attack. There were a number of proprietary IDSs on the market, but Snort's versatility and open-source licensing instantly appealed to the white hats, who loved nothing more than tinkering with a new security tool. Volunteer programmers jumped in to add functionality to the program.

Max was excited by Snort. The software was similar to BRO, the Lawrence Berkeley lab project that had helped sniff out Max's BIND attack, and Max knew it could be a game changer for online security. Now white hats could watch in real time for anyone trying to exploit the vulnerabilities discussed on Bugtraq and elsewhere. Snort was like an early-warning system for a network—the computer equivalent of the NORAD radar mesh that monitors America's airspace. All it was lacking was a comprehensive and up-to-date list of attack signatures, so the software would know what to look for.

In the first few months after Snort's release, a disorganized trickle of user-created signatures put the total number at about 200. In a single sleepless night, Max more than doubled the count, whipping up 490 signatures. Some were original, others were improved versions of the existing rules or ports from Dragon IDS, a popular proprietary system. Writing a rule meant identifying unique characteristics in the network traffic produced by a particular attack, like the port number or a string of bytes. For instance, the incantation `alert udp any any -> $INTERNAL 31337 (msg:"BackOrifice1-scan"; content:"|ce63 d1d2 16e7 13cf 38a5 a586|";)` detected black hats trying to use the Cult of the Dead Cow's Back Orifice malware that had so transfixed the crowd at Def Con 6.0. It told Snort that an incoming connection to port 31337, with a particular string of twelve bytes in the network traffic, was someone trying to exploit the back door.

Max put the signatures online as a single file on Whitehats.com, crediting a handful of other security geeks for their contributions, including Ghost23—a nod to his alter ego. Later, he converted the file to a full-fledged database and invited other experts to contribute their own rules. He gave the project the catchy name arachNIDS, for Advanced Reference Archive of Current Heuristics for Network Intrusion Detection Systems.

ArachNIDS was a hit and helped Snort surge to new levels of popularity in the security community, with Max Vision riding the swell to security stardom. As more white hats contributed to the project, it became the computer-security equivalent of the FBI's fingerprint database, capable of identifying virtually every known attack technique and variant. Max built on his success by writing papers dissecting Internet worms with the same clear eye he'd applied to the ADM worm. The technology press started seeking him out for comment on the latest attacks.

In 1999, Max injected himself into another promising venture aimed directly at tricking black-hat hackers. The Honeynet Project, as it would later be called, was the work of a former Army officer who applied his interest in military tactics to erect network "honeypots"—decoy computers that served no purpose but to be hacked. The Honeynet Project would secretly wire a packet sniffer to the system and place it unprotected on the Internet, like an undercover vice cop decked out in pumps and a short skirt on a street corner.

When a hacker targeted a honeypot, his every move would be recorded and then analyzed by security experts, with the results released to the world in the spirit of full disclosure. Max delved into the forensic work, reconstructing crimes from raw packet data and producing cogent analyses that blew the lid off some of the underground's concealed techniques.

But Max knew his rising recognition as a white hat wouldn't save him from the federal grand jury. In quiet moments, he fantasized with Kimi about escaping his fate. They could run off together, to Italy or some remote island. They'd start over. He'd find a benefactor, someone with money who recognized Max's talent and would pay him to hack.

The couple's relationship was suffering under the weight of the government's silent looming presence in their life. Before the raid, they hadn't much planned for the future. Now they couldn't. The future had been taken out of their control, and the uncertainty was toxic. They fought in private and snipped at each other in public. "The reason I signed the confession is because we'd just gotten married, and I didn't want to hurt you," Max said. He blamed himself, he added. By getting married, he'd given his enemies a weapon to use against him, a fatal flaw.

Kimi transferred from De Anza, a community college, to UC Berkeley, and the couple moved across the bay to live just off campus. The move proved fortuitous for Max. In the spring of 2000, a Berkeley company named Hiverworld offered him a long-awaited shot at the dot-com success that had already graced other Hungry Programmers. The company's plan was to create a new antihacking system that would detect intrusions, like Snort, but also actively scan the user's network for vulnerabilities, allowing it to ignore malicious volleys that had no chance of success. Snort author Marty Roesch was employee number 11. Now the company wanted Max Vision as number 21.

Max's first day was set for March 21. It was an early position at a promising technology start-up. The American dream, circa 2000.

On the morning of March 21, 2000, the FBI knocked on Max's door.

At first he thought it was a Hiverworld hazing, a practical joke. It wasn't. "Just don't answer it!" he said to Kimi. He grabbed a phone and found a hiding place, in case the agents peered through the windows. He dialed Granick and told her what was happening. The indictment must have finally come down. The FBI was there to take him to jail. What should he do?

The agents left—their arrest warrant didn't authorize them to crash into Max's home, so he'd temporarily thwarted them by the simple act of not answering the door. On her end, Granick called the prosecutor to try to arrange for a civilized self-surrender at the FBI field office in Oakland. Max contacted Hiverworld's CTO, his new boss, to report that he wouldn't be showing up for his first day at work. He'd be in touch in a day or two to explain everything, he said.

The evening news beat him to the punch: Alleged computer hacker Max Butler had just turned himself in on a fifteen-count indictment charging illegal interception of communications, computer intrusion, and possession of stolen passwords.

After two nights in jail, Max was brought in front of a federal magistrate in San Jose for arraignment. Kimi, Tim Spencer, and a dozen Hungry Programmers filled the gallery. Max was released on a $100,000 bond—Tim signed for half, and a fellow Hungry who'd struck it rich at a dot-com put down the remainder in cash.

The arrest sent shock waves through the computer security world. Hiverworld canceled its job offer on the spot—no security start-up could hire a man facing current computer intrusion charges. The community fretted over what would happen to the arachNIDS database without Max's curatorship. "It's his stuff," Roesch ruled in a post on a security mailing list. "So barring him explicitly ceding it to someone, it's still his to maintain."

Max responded personally in a long message sweeping through his early love of computers and the future direction of intrusion detection. Whitehats.com and arachNIDS would continue no matter what, he predicted. "My family and friends have been incredibly supportive and there are offers to maintain the sites to a certain degree should tragedy occur."

Casting himself as a victim, he railed against the "frenzy of the hacker witch-hunt" and slammed Hiverworld for disloyalty. "After the smoke cleared and I was in the press, Hiverworld decided not to continue our relationship," he wrote. "The corporation expressed cowardice that is deplorable. I can't tell you how disappointed I was to feel the complete lack of support from the Hive.

"I am innocent until proven guilty," he wrote. "And would appreciate the recognition of this by our community."

Six months later, Max pleaded guilty. The news was nearly lost amid a flurry of federal hacker prosecutions. The same month, Patrick "MostHateD" Gregory, the leader of a hacker gang called globalHell, was sentenced to twenty-six months in prison and ordered to pay $154,529.86 in restitution for a string of website defacements. At the same time, prosecutors charged twenty-year-old Jason "Shadow Knight" Diekman of California with cracking NASA and university systems for fun, and sixteen-year-old Jonathan James, known as "C0mrade," received a six-month sentence for his recreational intrusions into Pentagon and NASA computers—the first term of confinement ever handed down in a juvenile hacking case.

To all appearances, federal law enforcement now had firm control of the computer intrusions that had for so long struck fear into corporate America and government officials. In truth, all these victories were battles in yesterday's cyberwar against bedroom hackers, a dying breed. Even as Max copped his plea in a San Jose courtroom, the FBI was discovering a twenty-first-century threat gathering five thousand miles away—one intimately entwined with Max Vision's future.

8

Welcome to America

The two Russians made themselves at home in the small office in Seattle. Alexey Ivanov, twenty, typed on a computer keyboard while his associate, nineteen-year-old Vasiliy Gorshkov, stood by and watched. They were straight off a flight from Russia and already knee-deep into the biggest job interview of their lives—negotiating for a lucrative international partnership with the U.S. computer-security start-up Invita.

Office workers milled around them, and tinny pop music spilled from the computer's speaker. After a few minutes, Gorshkov drifted off to another computer across the room, and Michael Patterson, Invita's CEO, struck up a conversation.

It had been Patterson who'd invited the Russians to Seattle. Invita, he'd told them in an e-mail, was a young company, but it was gaining customers through contacts the founders had made while working at Microsoft and Sun. Now the company wanted help expanding into Eastern Europe. Ivanov, who claimed to have as many as twenty talented programmers working with him, seemed perfect for the job; Gorshkov was a tag-along, invited by Ivanov to act as the duo's spokesman. He had a fiancée waiting back home, pregnant with his first child.

Patterson began casually asking Gorshkov about a recent rash of computer intrusions into U.S. companies, some of whom paid money to the attackers to make them stop. "Just so I know you guys are as good as I think you are," Patterson said, "could any of that have been you guys?"

Gorshkov—bundled in the heavy jacket he wore back home in Chelyabinsk, a bleak, polluted industrial city in the Ural Mountains—hedged for a minute and finally answered. "A few months ago we tried, but we found it's not so profitable."

The Russian was being modest. For nearly a year, small to midsized Internet companies around the United States had been plagued by extortionate cyberattacks from a group calling itself the Expert Group of Protection Against Hackers—a name that probably sounds better in Russian. The crimes always unfolded the same way: Attackers from Russia or Ukraine breached the victim's network, stole credit card numbers or other data, then sent an e-mail or a fax to the company demanding payment to keep quiet about the intrusion and to fix the security holes the hackers exploited. If the company didn't pay up, the Expert Group would threaten to destroy the victim's systems.

The gang had lifted tens of thousands of credit card numbers from the Online Information Bureau, a financial transaction clearinghouse in Vernon, Connecticut. The Seattle ISP Speakeasy had been hit. Sterling Microsystems in Anaheim, California, had been hacked, along with a Cincinnati ISP, a Korean bank in Los Angeles, a financial services company in New Jersey, the electronic payment company E-Money in New York, and even the venerable Western Union, which had lost nearly sixteen thousand customer credit card numbers in an attack that came with a $50,000 extortion threat. When music-seller CD Universe didn't give in to a $100,000 ransom demand, thousands of its customers' credit card numbers showed up on a public website.

Several companies wound up paying the Expert Group small amounts to go away, while the FBI did its best to track the intrusions. They finally zeroed in on one of the ringleaders, "subbsta," whose real name was Alexey Ivanov. It wasn't that hard—the hacker, convinced he was out of reach of American justice, had given his résumé to Speakeasy during the extortion negotiations there.

Russian police had ignored a diplomatic request to detain and question Ivanov, and that was when the feds created Invita, a full-blown undercover business designed to lure the hacker into a trap. Now Ivanov and Gorshkov were surrounded by undercover FBI agents posing as company employees, along with a white-hat hacker from the nearby University of Washington who was playing the role of a computer geek named Ray. Hidden cameras and microphones recorded everything in the office, and FBI-installed spyware captured every keystroke typed on the computers. In the parking lot outside, around twenty FBI agents were standing by to help with the arrest.

The agent playing CEO Patterson tried to draw Gorshkov out some more. "What about credit cards? Credit card numbers? Anything like that?"

"When we're here, we'll never say that we got access to credit card numbers," the hacker replied.

The FBI agent and Gorshkov laughed conspiratorially. "I understand. I hear ya, I hear ya," said Patterson.

When the two-hour meeting concluded, Patterson ushered the men into a car, ostensibly to take them to the temporary housing arranged for their visit. After a short drive, the car stopped. Agents threw open the doors and arrested the Russians.

Back at the office, an FBI agent realized the keystroke logger installed on the bureau computers at Invita presented him with a rare opportunity. What he did next would make him the first FBI agent to be accused by the Russian federal police of committing a computer crime. He went into the keystroke logs and retrieved the password the pair had used to access their computer in Chelyabinsk. Then, after checking with his supervisor and a federal prosecutor, he logged in to the hackers' Russian server over the Internet and started scrounging through the directory names, looking for the files belonging to Ivanov and Gorshkov.

When he found them, he downloaded 2.3 gigabytes of compressed data and burned it onto CD-ROMs, only later obtaining a warrant from a federal judge to search through the information he'd grabbed. It was the first international evidence seizure through hacking.

When the feds dug into the data, the breathtaking scope of Ivanov's activity became clear. In addition to the extortion plots, Ivanov had developed a frighteningly effective method for cashing out the cards he stole, using custom software to automatically open PayPal and eBay accounts and bid on auctioned goods with one of the half-million stolen credit cards in his collection. When the program won an auction, it had the goods shipped to Eastern Europe, where an associate of Ivanov picked them up. Then the software did it all again and again. PayPal checked the stolen credit card list against its internal databases and found it had absorbed a stunning $800,000 in fraudulent charges.

It was the first tremor in a tectonic shift that would fundamentally change the Internet for the next decade. Maybe forever. With top-flight technical colleges but few legitimate opportunities for their graduates, Russia and the former Soviet satellite states were incubating a new breed of hacker.

Some, like Ivanov, were amassing personal fortunes by looting consumers and companies, protected by corrupt or lazy law enforcement in their home countries and poor international cooperation. Others, like Gorshkov, were driven into crime by tough economic circumstances. The hacker graduated from Chelyabinsk State Technical University with a degree in mechanical engineering and sank a small inheritance from his father into a computer-hosting and Web-design business. Despite his swaggering hacker machismo at Invita, Gorshkov had been a late addition to Ivanov's gang, and he'd paid his own way to America in the hope of improving his fortunes. In a way, he did: After his arrest in Seattle, he was earning more in prison doing janitorial and kitchen work at eleven cents an hour than his fiancée was drawing on public assistance back home.

After his arrest, Ivanov began cooperating with the FBI, rattling off a list of friends and accomplices still hacking back home. The bureau realized there were dozens of profit-oriented intruders and fraud artists from Eastern Europe already reaching their tentacles into Western computers.

In the years to come, the number would grow to thousands. Ivanov and Gorshkov were Magellan and Columbus: Their arrival in America instantly redrew the global cybercrime map for the FBI and placed Eastern Europe indisputably at its center.

9

Opportunities

Max wore a blazer and rumpled cargo pants to his sentencing hearing and watched silently as the lawyers sparred over his fate.

Jennifer Granick, the defense attorney, told Judge James Ware that Max deserved a lowered sentence for his service as the Equalizer. The prosecutor took the opposite position. Max, he argued, had *pretended* to be an FBI informant while secretly committing crimes against the U.S. government. It was worse than if he had never cooperated at all.

It was a strange sentencing hearing for a computer criminal. A dozen of Max's colleagues in the security world—people devoted to thwarting hackers—had written to Judge Ware on Max's behalf. Dragos Ruiu, a prominent security evangelist in Canada, called Max "a brilliant innovator in this field." French programmer Renaud Deraison credited Max's early support with making possible Nessus, Deraison's vulnerability scanner and one of the most important free security tools then available. "Given Max's potential and his clear vision of Internet security . . . it would be more useful for society as a whole that he stays among us as a computer security specialist . . . rather than spend time in a cell and see his computing talent go through a slow but sure decay."

From a technology worker in New Zealand: "Without the work that Max has done . . . it would be so much harder for my company and countless others to protect themselves from hackers." From a fan in Silicon Valley: "Taking Max out of the security community would greatly hurt our ability to protect ourselves." A former Defense Department worker wrote, "To imprison this individual would be a travesty."

Several of the Hungries wrote letters as well, as did Max's mother and sister. In her note, Kimi pleaded eloquently for Max's freedom. "He saved my life by helping me out of an abusive relationship and teaching me the meaning of self-respect," she wrote. "He gave me shelter when I had no place to live. He took very good care of me when I was seriously ill, saving my life again by taking me to the emergency room when I protested that I was 'fine' even as I was dying."

When the lawyers finished their arguments, Max spoke for himself, with the earnest politeness he always exhibited away from his computer. His attack, he explained, had been born of good intentions. He'd just wanted to close the BIND hole and had lost his head.

"I got swept up," he said softly. "It's hard to explain the feelings of someone who's gotten caught up in the computer security field. . . . I felt at the time that I was in a race. That if I went in and closed the holes quickly, I could do it before people with more malicious intentions could use them.

"What I did was reprehensible," Max continued. "I've hurt my reputation in the computer security field. I've hurt my family and friends."

Judge Ware listened attentively but had already made up his mind. Letting Max off without a prison term would send the wrong message to other hackers. "There's a need for those who would follow your footsteps to know that this can result in incarceration," the judge said.

The sentence: eighteen months in prison, followed by three years of supervised release in which Max wouldn't be allowed on the Internet without the permission of his probation officer.

The prosecutor asked the judge to order Max immediately taken into custody, but Ware denied the request and gave the hacker a month to put his affairs in order and turn himself in to the U.S. marshals.

. . .

Max and Kimi had moved to Vancouver, near her family, after his guilty plea. When they returned home, Max wasted no time arranging for Whitehats.com and arachNIDS to survive his incarceration. He set up automatic bill payments for his bandwidth and wrote out a list of items for Kimi to take care of in his absence. She was in charge of arachNIDS now, he said, indicating the server squatting on the floor of their apartment.

The couple adopted two kittens to keep Kimi company while he was gone, named for the swords from *Elric of Melniboné*. The orange boy-cat was Mournblade; the gray female was Stormbringer.

Max spent his last weekend of freedom in front of his keyboard, getting arachNIDS ready for Kimi's stewardship. When Monday came he turned himself in on schedule. On June 25, 2001, he was locked in the county jail pending his shipment to his new home, Taft Federal Prison, a corporate-run facility owned by Wackenhut, positioned near a small town in central California.

As far as Max was concerned, it was another injustice, just like back in Idaho. He'd been sent back to prison not for his hacking but for refusing to set up Matt Harrigan. He was being punished for his loyalty, once again a victim of a capricious justice system. He doubted Judge Ware had even looked at the details of his case.

Kimi was adrift, alone for the first time since she'd met Max. For all his talk about staying with her forever, he'd chosen a course of action that guaranteed their separation.

Two months later, Kimi was talking to him on the phone from prison when she heard a *pop!* and the smell of acrid smoke filled her nostrils. The motherboard on Max's server had burst into flames. Max tried to calm her—all she had to do was replace the motherboard. He could do it in his sleep. Max talked her through the process, but Kimi was realizing she wasn't cut out for life as the prison wife of a hacker.

In August, she went to the Burning Man festival in Nevada to forget her troubles. When she got home, she broke some bad news to Max over the phone. She'd met someone else.

It was another betrayal. Max took the news with eerie calm, interrogating her about every detail: What drugs was she on when she cheated on him? What sexual positions did they use? He wanted to hear her ask for his forgiveness—he'd have given it to her in a heartbeat. But that wasn't what she was asking for. She wanted a divorce. "I don't know if you even think about the future anymore," she said.

In search of closure, Kimi caught a flight to California and drove to Taft, where she sat nervously in the waiting room, her eyes playing over a wall of posters depicting Wackenhut's network of hivelike prisons around the country. When Max was brought in, he took his place across the stainless steel picnic table in the visiting room and launched into an appeal. He did think of the future, he told her, and he'd been making plans in the joint.

"I've been talking to some people," he said, lowering his voice to a hush. "People I think I could work with."

Jeffrey James Norminton was at the tail end of a twenty-seven-month stretch when Max met him in Taft. At thirty-four, Norminton had the stolid physical presence of a brawler, thick necked with an oversized forehead and a Kirk Douglas cleft in his chin. An alcoholic and an accomplished con man, he was a financial wizard who did his best work half-sober. He'd start chain-chugging Coors Lights as soon as he rolled out of bed, and by the end of the day he'd be useless, but in that sweet spot between the morning's sobriety and the blurriness of midafternoon, Norminton was a master of the high-stakes con—a criminal rainmaker who could produce seven-figure sums from thin air.

Norminton's latest caper had required little more than a telephone and a fax machine. The target had been the Entrust Group, a Pennsylvania investment brokerage house. On a summer day in 1997, Norminton picked up the phone and called a vice president at Entrust, adopting the persona of an investment manager at Highland Federal Bank, a real bank in Santa Monica, California.

Oozing confidence and charm, the swindler persuaded Entrust to buy into the bank's high-yield certificates of deposit, promising the VP a healthy 6.20 percent return on a one-year investment. When Entrust eagerly wired $297,000 to Highland, the cash wound up in the account of a dummy corporation Norminton's accomplice had set up under Entrust's name. To the bank, the transaction looked like an investment house moving money from one branch to another.

The grifters promptly withdrew all but $10,000 of the cash and then ran the scam again, this time with Norminton's partner making the phone call to the same VP and pretending to be from a different bank, City National, offering an even higher return. Entrust promptly sent two more transfers totaling $800,000.

Norminton was undone by his ambition. He sent his accomplice into City National to pull out $700,000 in a single cashier's check. An investigator at the bank got suspicious and backtracked the incoming wire transfers to the real Entrust. At the next withdrawal, FBI agents were waiting. The financial mastermind was now cooling his heels in Taft. The only silver lining to his incarceration was that he'd met a talented hacker looking to get back at the system.

Norminton made it clear that he saw real potential in Max, and the pair took to walking the yard every day, swapping war stories and fantasizing about how they might work together when they hit the streets. With Norminton's guidance, Max could easily learn to crack brokerage houses, where they'd tap into overstuffed trading accounts and drain them into offshore banks. One big haul and they'd have enough cash for the rest of their lives.

After five months, Norminton and his schemes were sent home to sunny Orange County, California, while Max remained at Taft with another year left on his sentence—long, tedious days of bad food, standing for count, and the sound of chains and keys.

In August 2002, Max was granted early release to a sixty-one-bed halfway house in Oakland, where he shared a room with five other ex-cons. Kimi met with Max to present him with divorce papers. She was getting serious with the guy she'd met at Burning Man; it was time, she said, for Max to let her go. Max refused to sign.

Max's relative freedom at the halfway house was tenuous—the facility demanded that he obtain gainful employment or go back to prison, and telecommuting wasn't allowed. He reached out to his old contacts in Silicon Valley and found his employability had been shattered by his high-profile hacking conviction and over a year in prison.

Desperate, he borrowed a laptop from one of the Hungry Programmers and banged out a message to an employment list watched by the computer security experts who had once admired him. "I have been showing up at places that farm out manual labor, 5:30 am, and still haven't found any work," he wrote. "My situation is just ridiculous." He offered his services at fire-sale prices. "I am willing to work for minimum wage for the next few months. Surely there is some open position at a security company in the area. . . . The last half dozen employers I have had paid me at least $100/hr for my time, now I am only asking for $6.75."

A consultant answered the plea, agreeing to let Max work out of his home office in Fremont, a short BART ride from the halfway house. He'd pay ten dollars an hour for Max to help build servers, a throwback to Max's first job for his father as a teen. Tim Spencer loaned Max a bike to pedal to the train station every day. Max was freed from the halfway house after two months, and the Hungry Programmers once again stepped up to provide him with shelter. He moved into an apartment in San Francisco shared by Chris Toshok, Seth Alves—a veteran of the Meridian master-key adventure—and Toshok's ex-girlfriend Charity Majors.

Despite the jailhouse fantasies he and Norminton had hatched, Max was determined to go straight. He resumed his search for work. But the job offers failed to pour in for the ex-con. Even the Honeynet Project, to which he'd donated his expertise just a couple of years earlier, shunned him.

His lot began improving in other ways: He started dating his housemate Charity Majors, a fellow Idaho refugee who designed herself like an avatar from a virtual world, painting her fingernails like Skittles—each a different color—and wearing contact lenses that tinted her eyes an impossible emerald. Money was tight for both of them: Charity worked as a system administrator for a porn website in Nevada, earning Silver State wages that were stretched thin in San Francisco. Max was nearly broke.

One of Max's former clients in Silicon Valley tried to help by giving Max a $5,000 contract to perform a penetration test on the company's network. The company liked Max and didn't really care if he produced a report, but the hacker took the gig seriously. He bashed at the company's firewalls for months, expecting one of the easy victories to which he'd grown accustomed as a white hat. But he was in for a surprise. The state of corporate security had improved while he was in the joint. He couldn't make a dent in the network of his only client. His 100 percent success record was cracking.

"I've never failed to get into a system before," Max told Charity in disbelief.

"Sweetie, you haven't touched a computer for years," she said. "It'll take you a little while. Don't feel like you have to get in today."

Max pushed harder, only becoming more frustrated over his powerlessness. Finally, he tried something new. Instead of looking for vulnerabilities in the company's hardened servers, he targeted some of the employees individually.

These "client side" attacks are what most people experience of hackers—a spam e-mail arrives in your in-box, with a link to what purports to be an electronic greeting card or a funny picture. The download is actually an executable program, and if you ignore the warning message on your Windows machine and install the software, your computer is no longer your own.

In 2003 the dirty secret of these attacks was that even savvy users who knew better than to install foreign software could be broadsided. "Browser bloat" was largely to blame. In the nineties a fierce battle with Netscape for control of the browser market had driven Microsoft to stuff Internet

Explorer with unnecessary features and functionality. Every added capability expanded the attack surface of the browser. More code meant more bugs.

Now Internet Explorer holes were constantly surfacing. They were usually discovered by one of the good guys first: Microsoft's own programmers or a white hat who often, but not always, warned the company before detailing the hole on Bugtraq.

But once a hole was public, the race was on. Black hats worked to exploit the bug by setting up Web pages serving the attack code and then tricking victims into visiting them. Just looking at the Web page would yield control of the victim's computer, without any outward sign of infection. Even if the bugs were not made public, the bad guys could figure them out by reverse-engineering the vulnerability from Microsoft's patches. Security experts had been watching with dismay as the time between a vulnerability's announcement and its exploitation by black hats shrank from months to days. In the worst-case scenario, the black hats found a bug first: a "zero day" vulnerability that left the good guys playing catch-up.

With new Microsoft patches coming out nearly every week, even vigilant corporations tended to lag in installing them, and average users often didn't patch at all. A global survey of one hundred thousand Internet Explorer users conducted around the time of Max's effort found that 45 percent suffered from unpatched remote access vulnerabilities; narrowing the field to American users cooled the number only slightly, to 36 percent.

Max's attack was effective. After securing access to an employee's Windows machine, he hopped on the company's network from the inside, grabbed some trophies, and popped out like the chest-bursting monster in *Alien*.

"It was then that I decided to scrap my old model of penetration testing and include client-centric attack as a mandatory part of the exercise," he later wrote a white-hat colleague. "I've been confident about the 100 percent rate ever since."

But instead of gratitude, Max's final report was greeted with outrage. Using a client-side attack in a penetration test was almost unseemly; if you were hired to test physical security at a company's corporate headquarters, you wouldn't necessarily feel free to burglarize an employee's home to steal the keys. The client gave him a tongue-lashing; they'd paid Max to attack their servers, not their employees.

Max began to wonder if he had a future in computer security at all. His former friends in the community had all moved on. Hiverworld, where Max had nearly been employee 21, revamped its executive team and won $11 million in venture capital, changing its name to nCircle Network Security. Marty Roesch left the company to build on the success of Snort—to which Max had contributed—starting a firm of his own called Sourcefire in Maryland. Both companies were on a path to success, nCircle kicking off an expansion that would take it to 160 employees in the years to come and Sourcefire heading to an IPO on the NASDAQ.

In some alternate universe in which Max had never hacked the Pentagon, or never used that Verio dialup, or had simply kept his mouth shut and worn a wire on Matt Harrigan, the hacker would have been riding one of those companies to financial success and rewarding, challenging work. Instead, he could only watch from the sidelines.

He was itinerant, grasping for cash, and flailing for something to do with his freedom. That was when he checked his Whitehats.com e-mail in-box and found an anonymous note from "an old friend from Shaft." It was the code phrase Max had worked out with Jeff Norminton.

Max met Jeff Norminton in a room at the St. Francis Hotel, and they caught up. Norminton hadn't taken well to supervised release: His sentencing judge required him to submit monthly urine samples, so his probation officer could make sure he hadn't started drinking again. That was a problem, since he was drinking again. After he'd refused two piss tests, the court had ordered him to check into Impact House, a drug and alcohol rehab center in Pasadena. He walked away after three weeks and was now looking to scam enough zeroes to flee to Mexico.

It was time to act on the plans they'd made in prison, Norminton said. He was ready to bankroll Max in his new career as a professional hacker.

Max was ready. He'd struggled long enough trying to make an honest living, and he was tired of being punished. He knew he was wearing out his welcome at the Hungry Programmers' house, even if they'd never complain. His diet was down to noodles and vegetables. He had no health insurance and dental problems that would cost thousands to fix.

Room service interrupted the conversation to deliver a hospitality basket. Norminton made a show of carrying the delivery into the bathroom, turning on the shower, and closing the door—in case the basket was bugged, he said. When they were done laughing, Max gave Norminton a short shopping list of gear he'd need to get started, a high-performance Alienware laptop, for one. And an antenna. A big one.

There was just one little hitch. Norminton was broke. They'd need to bring in someone else for seed money. Fortunately, Jeff knew just the guy.

10

Chris Aragon

Max met his future friend and criminal partner Chris Aragon in North Beach, San Francisco's little Italy, where seedy strip clubs and fortune tellers coexist with a row of pleasantly gaudy restaurants serving warm bread and hot pasta to sidewalk diners. The meeting was set for a coffee shop near the City Lights bookstore, cradle of the Beat Generation in the 1950s, and kitty-corner from Vesuvio Café, a saloon announced by colorful wall murals with wine bottles and a peace sign. Down the hill the Transamerica Pyramid stood sentry over the financial district, stabbing the sky.

Norminton introduced Chris to Max over the muted clatter of coffee cups and dishes. The two hit it off immediately. The forty-one-year-old Chris was a student of eastern spirituality, a vegetarian who practiced meditation to center his mind. Max, with his hippie values, seemed a kindred spirit on the road of life. They'd even read some of the same books.

And like Max, Chris had been arrested more than once.

It had all started in Colorado, when Chris was twenty-one years old. He was working as a masseuse at a hot springs resort, earning enough to cover his rent and support a modest cocaine habit, when he hooked up with a troubled veteran named Albert See whom he'd met in the joint while serving a juvenile sentence. See had just escaped from a minimum-security prison camp and needed money to get out of the country.

Chris came from a privileged background—his mother, Marlene Aragon, worked in Hollywood as voice talent, and she'd recently enjoyed a run on ABC's Saturday morning cartoon *Challenge of the Superfriends,* voicing Wonder Woman's feline nemesis the Cheetah. But he also had romantic notions of crime and criminals; on the wall of his condo hung a poster of the cover art from the Waylon Jennings album *Ladies Love Outlaws.* He took Albert in, and the two embarked on a series of bold, and mostly botched, bank robberies in the resort towns dotting Colorado.

The first robbery, at the Aspen Savings and Loan, started off well enough: Chris, wearing a blue and white bandana over his mouth to conceal his braces, pulled an Army-issue .45 automatic on the bank manager as he unlocked the door in the morning. He and Albert forced the manager inside, where they found a cleaning woman hiding under one of the desks, phoning the police. They left in a hurry.

The second robbery, at the Pitkin County Bank and Trust, was over before it even began. Chris's partner hid in a Dumpster by the back door, planning to jump out with his shotgun when the first employees came into work in the morning. The plan was aborted when Chris, watching from across the street, saw a garbage truck pull into the alley to empty the Dumpster.

The third robbery was better planned. On July 22, 1981, Chris and Albert visited Voit Chevrolet in Rifle and declared they wanted to test-drive a new Camaro. The luckless salesman insisted on going with them, and when they cleared the town limit, Chris steered to the side of the road, and Albert pulled the salesman from the car at gunpoint. They tied him up with rope, gagged him, and left him in a field before peeling away in the silver sports car.

The next day at 4:50 p.m., Chris drove the stolen Camaro up to the Valley Bank and Trust in Glenwood Springs, where the town locals parked the cash they earned from a flourishing tourist industry. Chris himself was a customer there. He waited outside behind the wheel of the car while Albert walked in wearing tinted sunglasses and toting a leather briefcase.

Albert ran out minutes later with $10,000 in cash and jumped into the Camaro, and Chris sped away.

Chris drove them south out of town on an unpaved road that snaked through the rocky red hills surrounding Glenwood Springs, then transferred to a jeep trail where his girlfriend was waiting with the switch car. Jubilant and excited, Chris drove past her and spun the Camaro into a triumphant fishtail, sending a plume of dust twenty feet into the air.

He was jumping up and down and shouting, "We did it!" when a police cruiser, drawn by the dust cloud, rolled up on the robbers. Chris and Albert made a mad dash on foot over the craggy, tree-dotted terrain. Chris tumbled down a ridge and landed on a cactus, and the two cops caught up with them. Chris dropped his shotgun and surrendered.

Chris learned a valuable lesson from his experience: not that crime didn't pay, but that guns and getaway cars were a stupid way to rob a bank. When he made parole in 1986, after five years in federal prison, he delved into credit card fraud and enjoyed some modest success. Then he hooked up with a Mexican drug smuggler he'd met in the joint. Chris helped with the delivery of two thousand pounds of marijuana to a twenty-acre ranch near Riverside, California, only to be busted in a nationwide DEA undercover operation. He went back to prison in September of 1991.

When he got out in 1996, he was thirty-five years old and had spent more than half his adult life, and a portion of his childhood, behind bars. He vowed to go straight. With his mother's help, he founded a legitimate business called Mission Pacific Capital, a leasing firm providing computer and business equipment to start-up companies hustling to claim their place in the dot-com race.

Clean-cut and handsome with an empathetic gaze, Chris fit easily into the role of a Southern California entrepreneur. After a lifetime of crime and uncertainty, the charms of a normal, middle-class existence had an exotic and satisfying appeal. He loved traveling to conventions, interviewing and hiring employees, schmoozing with colleagues. At a marketing convention in New Orleans, he met Clara Shao Yen Lee, a stylish woman of Chinese descent who'd emigrated from Brazil. Taken by Clara's beauty and intelligence, he promptly married her.

Under Chris's leadership, Mission Pacific built a reputation as an innovative leasing broker, one of the first to offer instant contracts through the Web, which helped the firm gain tens of thousands of clients around the country. The former bank robber and drug smuggler had two prominent Orange County businessmen as partners and twenty-one employees working in a spacious office a block from the Pacific Coast Highway. Clara dropped in periodically to help out with the look and feel of the company's website and marketing material. By 2000, the couple had an upscale condo in Newport Beach, a son, and had staked a claim in a business that seemed as limitless in its potential as the Internet itself.

That spring, the dream died; the dot-com bubble burst, and the torrent of new companies that had been Mission Pacific's lifeblood started to dry up. Then larger companies like American Express entered the leasing arena, squeezing out smaller firms. Chris's company was one of dozens of leasing brokers to crash and burn. He began shedding employees and finally had to tell the stragglers that Mission Pacific wouldn't be able to cut their next payroll checks.

Chris went to work for another leasing company but was cut in a round of layoffs when a large bank acquired the firm. Meanwhile, his wife gave birth to a second boy. So when Jeff Norminton showed up talking about the superhacker he'd met in Taft, Chris was ready to listen.

By the time he and Max met in that North Beach restaurant, Chris had already been funding Norminton's scheme, providing some of the specialized equipment Norminton said his hacker needed. Now that Chris had met Max in person, he was eager for a demonstration. After talking for hours, the three of them left the coffee shop to find someplace to hack from.

They wound up at the twenty-seven-story Holiday Inn in Chinatown, a few blocks away. At Max's direction, they asked for a room high above the street. Max positioned himself at the window, booted his laptop, plugged in the antenna, and began scanning for Wi-Fi networks.

In 2003, the world was going wireless in a big way and bringing a massive security hole with it. The revolution had begun with Apple's AirPort wireless access point and then was joined by hardware makers like Linksys and Netgear. As hardware prices dropped, more and more companies and home users began breaking free of the tethers of their blue Ethernet cables.

But the wireless gear being ushered into homes and offices around the country was a hacker's dream. It overwhelmingly employed a wireless standard called 802.11b, which included an encryption scheme that, in theory, would make it difficult to jump onto someone's wireless network without authorization or to passively eavesdrop on computer traffic. But in 2001, researchers at the University of California at Berkeley revealed a number of severe weaknesses in the encryption scheme that made it crackable with ordinary off-the-shelf equipment and the right software. And as a practical matter that technical black magic was usually not even needed. To speed adoption, manufacturers were shipping wireless access points with encryption turned off by default. Businesses small and large simply plugged in the boxes and forgot about them—sometimes assuming falsely that their office walls would keep their networks from seeping out onto the street.

A few months before Max went to jail, a white-hat hacker had invented a sport called "war driving" to highlight the prevalence of leaky networks in San Francisco. After slapping a magnetically mounted antenna to the roof of his Saturn, the white hat cruised the city's downtown streets while his laptop scanned for beaconing Wi-Fi access points. After one hour in the financial district, his setup would find close to eighty networks. A year and a half had passed since then, and San Francisco, like other large cities, was now blanketed in an invisible sea of network traffic, available to anyone who cared to dip in.

Hacking from home was for idiots and teenagers—Max had learned that lesson the hard way. Thanks to Wi-Fi, he could now work from almost anywhere with complete anonymity. This time, if the police traced back one of Max's hack attacks, they'd wind up on the doorstep of whatever poor sap Max had used for connectivity.

The antenna Max used was a monster, a two-foot-wide wire-grid parabolic that quickly teased out dozens of networks from the ether surrounding the Holiday Inn. He jumped on one and showed Chris how it all worked. Wielding a vulnerability scanner—the same kind of tool he'd used in his pen tests—he could quickly scan huge chunks of Internet address space for known vulnerabilities, like sending a drift net into the Web. Security holes were everywhere. He was confident he'd be in financial institutions and e-commerce sites in no time. It was up to Norminton and Chris to decide what kind of data they needed and how they'd exploit it.

Chris was blown away. This six-foot-five, semi-vegetarian hacker knew his stuff, even if he was rusty from the joint.

Chris introduced Max to one of his prison contacts, a real estate fraudster named Werner Janer whom Chris had met in Terminal Island in '92. Janer offered to pay Max $5,000 to penetrate the computer of a personal enemy. He wrote the check out to Charity so Max wouldn't have to explain the income to his probation officer.

The money gave Max some breathing room. He began flying to Orange County, misspelling his name on the ticket so there'd be no record of his violating his supervised release by leaving the Bay Area. He and Norminton began crashing at Chris's place for a week at a stretch, hacking from Chris's garage.

He downloaded a list of small-sized financial institutions from the FDIC's website, figuring they'd be most vulnerable, and launched a script to scan each bank for known security holes. An electronic chime rang out through the garage whenever it scored a hit. He wormed into the banks and pulled out customer names, financial data, and checking account numbers.

The scattershot approach meant Max would be spared the frustration he'd felt in his last legitimate penetration test. Hacking any one particular target can be difficult; depending on the target, maybe even impossible. But scan hundreds or thousands of systems, and you're guaranteed to find some that are soft. It was a numbers game, like trying car doors as you walk through a parking lot.

Charity had only the broadest notion of what Max was up to, and she didn't like it. In an effort to win her over, Chris and Norminton invited the couple down to Orange County for a short vacation, paying their way for a weekend at Disneyland. Charity could see that Max and Chris were clicking, but something about Chris didn't smell right. He was too slick, too polished.

Max's hacking moved to small e-commerce sites, where he grabbed transaction histories, some with credit card numbers. But his efforts were unfocused, and neither Chris nor Norminton was sure what to do with all the data he was stealing.

Fortunately, Chris had some money coming in. Werner Janer owed him $50,000 and was ready to wire-transfer the money to a bank account of Chris's choosing. Determined to get his hands on cold, hard, unreported cash, Chris asked Norminton to do what he did best; Norminton agreed to have one of his friends receive the transfer and pull it out over the course of a few days.

The first round of withdrawals went as planned, and Norminton and his friend showed up at Chris's and handed over $30,000 in $100 bills. The following day, though, Norminton reported that his friend had taken ill and would have to take the day off.

In truth, Norminton had discovered the source of the windfall: It was Chris's cut from a real estate scam he'd helped Janer pull off. The money was dirty, and Norminton was now implicated in the scheme. The next morning, Chris found the Honda he'd loaned Norminton parked outside his office, one tire flat and a fresh dent in the fender. There was a note from Norminton inside: The FBI is after me. I'm skipping town.

Chris phoned Norminton's cash mule, already knowing what the score would be: Norminton's associate was in perfect health and had withdrawn the other $20,000 the day before, as planned. He'd given it to Norminton. Didn't Chris get it?

Chris tracked down Max through Charity and demanded answers: What did Max know about Norminton's whereabouts? Where was Chris's money? Max was as surprised as Chris at Norminton's disappearance, and eventually the two agreed to continue their partnership without Norminton.

Max and Chris fell into a routine. Once a month, Chris flew or drove north and met Max in downtown San Francisco, where they checked into a hotel. They'd carry Max's massive antenna up the fire stairs to their room and mount it on a tripod near the window. Then Max would putter for a while to locate a high-speed Wi-Fi with a strong signal.

They learned that altitude wasn't as important in Wi-Fi hacking as the sprawl of buildings visible out the window. If they came up dry, Chris would run down to the front desk to ask for a different room, explaining earnestly that he couldn't get a cell phone signal or was too afraid of heights to remain on the twentieth floor.

Max treated it like a job, saying good-bye to Charity and then vanishing for up to a week into one of the city's finest hotels, the Hilton, Westin, W, or Hyatt. While the clang of cable car bells rose from the streets below, Max cast his net over cyberspace, scooping up whatever data he could find—not really sure what he was looking for.

On a whim, he cracked Kimi's computer and that of her boyfriend, with whom she'd moved in. Max contemplated plundering her address book and sending out a mass e-mail in her name, detailing how she betrayed him. He thought everyone should know that Kimi's new life was built on a foundation of infidelity.

He didn't go through with it. He had Charity now. Kimi had moved on, and nothing would be gained by trying to shame her, he realized. Shortly thereafter, he signed the divorce papers.

Returning to his work, he began performing Google searches for guidance in his targeting: What were other fraudsters doing? How were they monetizing stolen data? That was when he discovered where the real criminal action was online: two websites called CarderPlanet and Shadowcrew.

11

Script's Twenty-Dollar Dumps

In the spring of 2001, some 150 Russian-speaking computer criminals convened a summit at a restaurant in the Ukraine port city of Odessa to brainstorm the launch of a revolutionary website. Present were Roman Vega, a thirty-seven-year-old man who sold counterfeit credit cards to the underground through his online storefront BOA Factory; a cybercrook known as "King Arthur"; and the man who would emerge as their leader, a Ukrainian credit card seller known by the handle "Script."

The discussion was sparked by the success of a UK-hosted website erected in 2000 called Counterfeit Library, which solved one of the fundamental weaknesses of conducting criminal business in IRC chat rooms, where the wisdom and experience of years of crime vanished into the air as soon as the chat was over. Founded by a handful of Western cybercrooks, Counterfeit Library collected underground tutorials onto a single website and attached an online discussion forum where identity thieves could gather to swap tips and buy and sell "novelty" identification cards—a euphemism distilled from the same spirit in which hookers go on "dates."

Counterfeit Library had more in common with the electronic bulletin board systems of the pre-Web days than with IRC. Members could post in permanent discussion threads and build personal reputations and brands. As criminals around the globe discovered this patch of dry land in the murky ephemeral sea of underground commerce, the site collected hundreds, then thousands, of members from across North America and Europe. They were identity thieves, hackers, phishers, spammers, currency counterfeiters, credit card forgers, all of whom had been slaving away in their apartments and warehouses, blind, until now, to the vastness of their secret brotherhood.

The carders of Eastern Europe had watched Counterfeit Library with envy. Now they wanted to apply the same alchemy to their own underground.

In June 2001, the result of the Odessa summit was unveiled: the International Carders Alliance, or simply Carderplanet.com, a tightly organized reinvention of Counterfeit Library catering to the underworld of the former Soviet empire. While Counterfeit Library was a freewheeling discussion board and BOA Factory a straightforward storefront operation, CarderPlanet was a disciplined online bazaar, charged with the excitement of a commodities exchange.

Unabashed in its purpose, the site adopted the nomenclature of the Italian Mafia for its rigid hierarchy. A registered user was a "sgarrista"—a soldier, without special privileges. One step up was a "giovane d'honore," who helped moderate the discussions under the supervision of a "capo." At the top of the food chain was CarderPlanet's don, Script.

Russian-speaking vendors flocked to the new site to offer an array of products and services. Credit card numbers were a staple, naturally, but only the beginning. Some sellers specialized in the more valuable "full infos"—a credit card number accompanied by the owner's name, address, Social Security number, and mother's maiden name, all for around $30. Hacked eBay accounts were worth $20. Ambitious buyers could spend $100 for a "change of billing," or COB, a stolen credit card account where the billing address could be changed to a mail drop under the buyer's control. Other vendors sold counterfeit checks or money orders, or rented drop addresses in the United States where merchandise ordered on American credit cards could be delivered without raising alarms and then reshipped to the scammer.

Physical products like blank plastic "magstripe" (magnetic stripe) cards were in the offering, as well as "novelty" IDs, complete with holograms, which sold for anywhere from $75 to $150, depending on the quality. One could purchase a package of ten identification cards with the same photo but different names for $500.

CarderPlanet's registration was open to anyone, but to sell on the site, vendors first had to submit their products or services to an approved reviewer for inspection. New vendors would sometimes be required to escrow their transactions through Script or to post a bond with the site's emergency fund, used to pay out buyers in case an approved vendor went out of business with unfilled orders in his queue. Vendors were expected to keep the board apprised of any vacation plans, safeguard buyers' information from hacker attacks, and respond promptly to customer complaints. "Rippers," vendors who failed to deliver on a sale, were subject to banishment, as was any vendor who accumulated five customer complaints.

CarderPlanet was soon imitated by a second site, this one aimed at the English-speaking world: Shadowcrew. In September 2002, after witnessing the stunning success of CarderPlanet's regimented hierarchy, a carder named "Kidd" brought over the heaviest hitters from Counterfeit Library to do business the Russian way. News of the site spread through IRC chat rooms and prison yards alike, and by April 2003, Shadowcrew had four thousand registered users.

With the motto "For Those Who Like to Play in the Shadows," Shadowcrew was at once a study-at-home college and an online supermarket for nearly anything illegal. Its tutorials offered lessons on how to use a stolen credit card number, forge a driver's license, defeat a burglar alarm, or silence a gun. It boasted a wiki that tracked which state driver's licenses were forgeable. And its approved vendors around the world could provide a dizzying array of illicit products and services: credit reports, hacked online bank accounts, and names, birth dates, and Social Security numbers of potential identity theft targets.

As on CarderPlanet, each product had its own specialists, and every vendor had to be reviewed by a trusted site member before they were allowed to sell. Disputes were handled judiciously, with administrators and moderators working overtime to expose and ban rippers selling bunk products.

The trading wandered beyond data into tangible items like ATM skimmers, prescription drugs, and cocaine, and into services like distributed denial-of-service (DDoS) attacks—take down any website for $200—and malware customization to evade antivirus products. One well-reviewed vendor offered a test-taking service that promised to get customers technical certifications within days. A vendor called UBuyWeRush sprang up to flood the underground with magnetic stripe writers, as well as must-haves like safety paper and magnetic ink cartridges for counterfeiting checks.

Child porn was forbidden, and one vendor who asked to be reviewed for exotic animal sales was laughed off the board. But nearly anything else was fair game on Shadowcrew.

By this time, CarderPlanet had launched subforums for criminals from Asia, Europe, and the States, but it was Shadowcrew that forged a true international marketplace: a cross between the Chicago Mercantile Exchange and *Star Wars*'s Mos Eisley cantina, where criminals of varying disciplines could meet up and collaborate on heists. An identity thief in Denver could buy credit card numbers from a hacker in Moscow, send them to Shanghai to be turned into counterfeit cards, then pick up a fake driver's license from a forger in Ukraine before hitting the mall.

Max shared his discovery with Chris, who was fascinated. Chris logged on to the forums and studied the content like a textbook. A lot of things hadn't changed since he'd dealt in credit card fraud in the 1980s. Other things had changed a lot.

There was a time when crooks could literally pull credit card numbers from the trash by Dumpster-diving for receipts or the carbon-paper slips left over from retailers' sliding imprint machines. Now mechanical imprinting was dead, and Visa and MasterCard insisted that receipts not include full credit card account numbers. Even if you got the numbers, that was no longer enough to make counterfeit cards. The credit card companies now added a special code to every magnetic stripe—like a PIN, but unknown even to the cardholder.

Called a Card Verification Value, or CVV, the code is a number distilled from other data on the stripe—primarily the account number and expiration date—and then encrypted with a secret key known only to the issuing bank. When the magstripe is swiped at the point-of-sale terminal the CVV is sent along with the account number and other data to the issuing bank for verification; if it doesn't match, the transaction is declined.

When it was introduced by Visa in 1992, the CVV began driving down fraud costs immediately, from nearly .18 percent of Visa transactions that year to around .15 percent a year later. In the 2000s, the innovation proved a strong bulwark against phishing attacks, in which a spammer spews thousands of falsified e-mails aimed at luring consumers into entering their credit card numbers into a fake bank website. Without the CVV on the magnetic stripe—which consumers didn't know, and thus couldn't reveal—those stolen numbers were useless at real-world cash registers. Nobody could walk into a Vegas casino, slap down a card derived from a phishing attack, and get a pile of black chips to carry to the roulette table.

MasterCard followed Visa's lead with its own Card Security Code, or CSC. Then in 1998, Visa introduced the CVV2, a different secret code printed on the backs of cards for consumers to use exclusively over the phone or the Web. That further reduced crime losses and completed the Chinese wall between fraud on the Internet and in real life: Accounts stolen from e-commerce sites or in phishing attacks could only be used online or over the phone, while magstripe data could be used in-store but not on the Web, because it didn't include the printed CVV2.

By 2002, the security measure had turned raw magstripe data into one of the underground's most valuable commodities and pushed the point of compromise closer to the consumer.

Hackers began breaching transaction-processing systems for the data, but the most straightforward way for ordinary crooks to steal the information was to recruit a cash-hungry restaurant employee and equip him with a pocket-sized "skimmer," a magstripe reader with built-in memory. As small as a cigarette lighter and readily concealed in the apron pocket of a fast-food worker or the suit jacket of an upscale maître d', a skimmer can hold hundreds of cards in its memory for later retrieval through a USB port. A server needs only a second of privacy to swipe a customer's card through the device.

In the late 1990s, thieves began fanning out in big cities across the United States, eyeing waiters, waitresses, and drive-through attendants who might be interested in a little extra cash, typically $10 a swipe. Though it was riskier, gas station managers and retail workers could get in on the action as well by installing tiny skimming circuit boards in pay-at-the-pump readers and point-of-sale terminals. Some of the data would be exploited locally, but much of it was sent to Eastern Europe, where the swipes were sold over the Internet ten, twenty, a hundred, or even thousands at a time.

The carders call these "dumps"; each contained just two lines of text, one for each track on a credit card's three-inch-long magstripe.

```
Track 1: B4267841463924615^SMITH/
JEFFREY^04101012735200521000000
Track 2: 4267841463924615=041010127352521
```

A dump was worth about $20 for a standard card, $50 for a gold card, and $80 to $100 for a high-limit corporate card.

Chris decided to try some carding himself. He determined that Script, the godfather of CarderPlanet, was the most reliable source of dumps in the world. He paid the Ukrainian $800 for a set of twenty Visa Classic numbers and elsewhere parted with around $500 for an MSR206, the underground's favorite magnetic stripe encoder.

Once the shoebox-sized MSR206 was plugged into his computer and the right software installed, he could take an anonymous Visa gift card, or one of his own credit cards, and encode it in two quick swipes with one of Script's dumps.

With the reprogrammed card burning a hole in his pocket, Chris browsed his local Blockbuster and some retailers to scope out the opportunities. Simple magstripe fraud might be cheap and easy, but it had severe limitations. Through observation, Chris quickly determined that shopping for consumer electronics or expensive clothes would be tough: To guard against what Chris was contemplating, many high-end stores require the checkout clerk to physically type the last four digits from the face of the credit card; the point-of-sale terminal rejects the card, or worse, if the digits don't match what's on the stripe. A reprogrammed card was only good at spots where employees never get to lay their hands on the plastic, like gas stations or drugstores.

Chris made his move at a local supermarket. He loaded his cart indiscriminately and checked out, sliding his plastic through the point-of-sale terminal. After a moment, the word "Approved" flickered across the display, and somewhere in America a random consumer was charged for $400 in groceries.

Chris delivered his ill-gotten groceries to an Orange County couple in worse financial shape than himself and then took the husband—a contractor who'd recently had his tools stolen—to a local Walmart to purchase new construction gear. Word spread that Chris had credit cards, and he began doling out his reprogrammed plastic to a few friends, who were always thoughtful enough to make small purchases for Chris as a thank-you.

He could see the outlines of a business plan in his circulating plastic. Drop everything else, he told Max. The real money is in dumps.

12

Free Amex!

Max broached his plan obliquely with Charity over the rare indulgence of a sushi dinner. "Which institutions would you say deserve to be punished the most?" he asked.

He had the answer ready: the moneylenders. The greedy banks and credit card companies who saddle consumers with $400 billion in debt each year while charging usurious interest and hooking kids on plastic before they've graduated college. And because consumers were never held directly liable for fraudulent charges—by law they could only be billed for the first $50, and most banks waived even that—credit card fraud was a victimless crime, costing only these soulless institutions money.

Credit wasn't real, Max reasoned, just an abstract concept; he would be stealing numbers in a system, not dollars in someone's pocket. The financial institutions would be left holding the bag, and they deserved it.

Charity had learned to accept the bitterness Max brought back from prison: Living with him meant never again watching a crime drama on TV, because any depiction of the police as good guys set Max fuming. She wasn't entirely sure what Max had in mind now, and she didn't want to know. But one thing was clear. Max had decided he was going to be Robin Hood.

. . .

Max knew exactly where to get the magstripe data Chris wanted. There were thousands of potential sources sitting in plain sight, right on CarderPlanet and Shadowcrew. The carders themselves would be his prey.

Most of them weren't hackers, they were just crooks; they knew a bit about fraud but little about computer security. They certainly wouldn't be much harder to hack than the Pentagon. It was also a morally palatable proposition: He would be stealing credit card numbers that had already been stolen—a criminal was going to use them, so it might as well be Chris Aragon, his criminal.

He started by choosing his weapon, picking out the slick Bifrost Trojan horse program already circulating online and customizing it to evade antivirus detection. To test the results, he used the computer emulation software VMware to run a dozen different virtual Windows boxes on his computer at once, each loaded with a different flavor of security software.

When the malware went undetected on all, he moved to the next step: harvesting a list of carders' ICQ numbers and e-mail addresses from public forum posts, collecting thousands of them into a database. Then, posing as a well-known dumps vendor named Hummer911, he fired off a message to the entire list. The note announced that Hummer911 had acquired more American Express dumps than he could use or sell, so he was giving some away. Click here, Max wrote, to get your free Amex.

When a carder clicked on the link, he found himself looking at a list of fake Amex dumps Max had generated, while invisible code on the Web page exploited a new Internet Explorer vulnerability.

The exploit took advantage of the fact that Internet Explorer can process more than just Web pages. In 1999, Microsoft added support for a new type of file called an HTML Application—a file written in the same markup and scripting languages used by websites but permitted to do things on a user's computer that a website would never be allowed to do, like creating or deleting files at will and executing arbitrary commands. The idea was to let developers already accustomed to programming for the Web use the same skills to craft fully functional desktop applications.

Internet Explorer recognizes that HTML Applications can be deadly and won't execute them from the Web, only from the user's hard drive. In theory.

In practice, Microsoft had left a hole in the way the browser screened content embedded on a Web page. Many Web pages contain OBJECT tags, which are simple instructions that tell the browser to grab something from another Web address—typically a movie or music file—and include it as part of the page. But it turned out you could also load an HTML Application through the OBJECT tag and get it to execute. You just had to disguise it a little.

While Max's victims salivated over the bogus American Express dumps, an unseen OBJECT tag instructed their browsers to pull in a malicious HTML Application that Max had coded for the occasion. Crucially, Max had given the file a name ending in ".txt"—a superficial indication that it was an ordinary text file. Internet Explorer saw that file name and decided it was safe to run.

Once the browser started downloading the file, however, Max's server transmitted a content type indicator of "application/hta"—identifying it now as an HTML Application. Essentially, Max's server changed its story, presenting the file as a harmless document for the browser's security check, then correctly identifying it as an HTML Application when it came time for the browser to decide how to interpret the file.

Having judged the file safe based on the name, Internet Explorer didn't reevaluate that conclusion once it learned the truth. It just ran Max's code as an HTML Application instead of a Web page.

Max's HTML Application was a tight Visual Basic script that wrote out and executed a small grappling-hook program on the user's machine. Max named the grappling hook "hope.exe." Hope was Charity's middle name.

The grappling hook, in turn, downloaded and installed his modified Bifrost Trojan horse. And just like that, Max was in control.

. . .

The carders converged like hungry piranhas on his poisoned page: Hundreds of their machines reported back to Max for duty. Excited, he began poking around the criminals' hard drives at random. He was surprised by how small-time it all looked. Most of his victims were buying small batches of dumps, ten or twenty at a time—even less. But there were lots of carders, and there was nothing to keep him from returning to their machines over and over again. In the end, the Free Amex attack would score him about ten thousand dumps.

He siphoned the dumps to Chris as he found them and vacuumed other useful data from his victims: details on their scams, stolen identity information, passwords, mailing lists used in phishing schemes, some real names, photos, and e-mail and ICQ addresses of their friends—useful for future attacks on the underground.

With a single well-constructed ruse, he was now invisibly embedded in the carders' ecosystem. This was the start of something big. He'd be a stick-up man among the carders, living off whatever he could skim from their illegal economy. His victims couldn't call the cops, and with his anonymous Internet connection and other precautions, he'd be immune to reprisal.

It wasn't long, though, before Max discovered that not all of the carders were what they seemed to be.

The victim was in Santa Ana. When Max strolled into the computer through his back door and began poking around, he saw at once that something was very wrong.

The computer was running a program called Camtasia that keeps a video record of everything crossing the computer's screen—not the kind of information a criminal normally wants to archive. Max foraged through the hard drive, and his suspicions were confirmed: The disk was packed with FBI reports.

Chris was shaken by the discovery of an FBI cybercrime agent in his own backyard, but Max was intrigued—the agent's hard drive offered potentially useful insight into the bureau's methods. They talked about what to do next. Some of the files indicated the agent had an informant who was providing information on Script, the CarderPlanet leader who sold Chris his first dumps. Should they warn Script that there was an informant in his circle?

They decided to do nothing; if he were ever busted, Max figured, he might be able to play this as a trump card. If it got out that he'd accidentally hacked an FBI agent, it could embarrass the bureau, maybe even cost them some convictions.

He returned to his work hacking the carders. But he knew now that he wasn't the only outsider worming into the crime forums.

13

Villa Siena

Palm trees rose at the entrance of Villa Siena, a sprawling gated community in Irvine, half a mile from John Wayne Airport. Beyond the front gate, European-inspired fountains bubbled in the manicured courtyards, and four swimming pools sparkled blue beneath the sunny Southern California sky. Residents were enjoying the clubhouse, relaxing in the spas, getting in a workout at one of the three fitness rooms, or perhaps visiting the full-time concierge to make plans for the evening.

In one of the spacious apartments, Chris Aragon was running his factory. The drapes were drawn over the giant picture window to hide the riot of machinery crowding the Ikea tables and granite countertops. He flipped on his card printer, and it awakened with a whining rumble, wheels spinning up to speed, motors pulling the ribbons taut as a hospital bedsheet.

Max was snagging dumps regularly now, and when he got a new haul, there was no time to waste—the swipes were stolen property twice over, and Chris had to burn through them before the crooks who'd purchased or hacked the numbers maxed them out first or blundered and got them flagged by the credit card companies. Chris had tapped the last of his reserves to invest in about $15,000 worth of credit card printing gear and the apartment to house it. Now the investment was paying dividends.

Chris loaded blank PVC cards into the hopper of an unwieldy oblong machine called a Fargo HDP600 card printer, a $5,000 device used to print corporate ID cards. With a click on his laptop, the machine drew a card into its maw and hummed once, twice, a third, and a fourth time, each sound marking another color as it moved to a clear transfer ribbon and was rapidly vaporized by heating elements and fused to the surface of the card. A final low grinding from the Fargo meant a clear laminate coat was settling over the plastic.

It was forty-four seconds from start to finish, and then the machine spat out the card—a glossy, brightly colored consumer objet d'art. A bald eagle staring purposefully at a Capitol One logo, or the grim American Express centurion, or the simple smudge of sky blue across the white face of a Sony-branded MasterCard. For the high-limit cards, the process was the same, except sometimes Chris would start with gold- or platinum-colored PVC stock, purchased, like the white cards, in boxes of hundreds.

Once he had a pile of freshly printed plastic in hand, Chris moved to a second stop in the assembly line: a monochrome printer for the fine print on the back of the card. Then if the design called for a hologram, he'd pluck a sheet of Chinese-produced counterfeits from a stack, align it carefully in a die punch, and pull the lever to cut out an oval or rounded rectangle the size of a postage stamp. A $2,000 Kwikprint Model 55 heat stamper, resembling a drill press crossed with a medieval torture instrument, fused the metal foil to the surface of the PVC.

The embosser was next: a giant motorized carousel wheel of letters and numbers that sounded like an IBM Selectric as it banged the name, account number, and expiration date one character at a time into the plastic, tipping each with silver or gold foil. From a Chinese supplier, Chris had obtained the special security keys for Visa's "flying V" and MasterCard's joined "MC"—two distinctive raised characters found only on credit cards, real and fake.

Credit card verification systems don't check the customer's name, which meant Chris had the luxury of choosing whatever moniker he liked for the front of his plastic; he preferred "Chris Anderson" for the cards he used himself. On his computer, Chris edited Max's dumps to make the name on the magstripe match the alias—conveniently, the name was the one piece of magstripe data not used in calculating the CVV security code, so it could be altered at will.

Finally, it was two swipes through the trusty MSR206 to program Max's dump onto the magstripe, and Chris had a counterfeit credit card that duplicated in nearly every way the plastic nestled in a consumer's wallet or purse somewhere in America.

He wasn't done yet.

Driver's licenses were a must for high-end purchases, and there, too, Chris's assembly line and Shadowcrew's tutorials got the job done. For licenses, he'd switch from PVC to Teslin, a thinner, more flexible material sold in 8½ x 11 inch sheets. It was one sheet for the front, another for the back, ten licenses to a sheet.

California licenses include two security features that took some extra hacking. One is a translucent image of the California state seal, set in a repeating pattern in the clear laminate over the face of the license. To simulate it, Chris used Pearl Ex, a fine colored powder sold at arts-and-crafts stores for less than three dollars a jar. The trick was to dust a sheet of laminate with a mix of gold and silver Pearl Ex, feed it into a printer loaded with a clear ink cartridge, and print a mirror image of the California pattern with the transparent ink. It didn't matter that the ink was invisible—it was the heat from the print head he was after. When the sheet came out, the printer had heat-fused the pattern onto the surface, and the extra Pearl Ex was easily washed away in a cold rinse.

The ultraviolet printing on the face of the license was no more difficult. An ordinary ink-jet printer would do the trick, as long as one drained the ink from the cartridge reservoirs and replaced it with multicolored UV ink bought in tubes.

After all the dusting, printing, and washing, Chris was left with four sheets of material. He would sandwich the two sheets of printed Teslin between the laminate and run it through a pressure laminator. After die-cutting, the result was impressive: Run your fingers over the license and feel the flawless silken surface; hold it at an angle and witness the ghostly state seal; put it under a UV bulb, and the state flag glowed eerily, the words "California Republic" in red, above them a brown bear walking on four legs across a yellow hilltop.

With cards and licenses complete, Chris got on the phone and summoned his girls. He'd figured out that attractive college-aged women made the best cashers. There was Nancy, a five-foot-three-inch Latina with "love" tattooed on one wrist; Lindsey, a pale girl with brown hair and hazel eyes; Adrian, a young Italian woman; and Jamie, who'd worked as a waitress at the Hooters in Newport Beach.

He'd met the twin brunettes Liz and Michelle Esquere at Villa Siena, where they lived. Michelle was just hanging around with the group, but Liz was invaluable: She had worked in the mortgage industry and was whip-smart, well educated, and responsible enough to take over some of the administrative work, like maintaining the spreadsheet of payouts, in addition to making in-store purchases.

Chris had a talent for recruitment. He might meet a new prospect at a restaurant and invite her to go partying with his friends. She'd join them at the clubs and expensive dinners, ride in the back of the rented limousine when one of them had a birthday to celebrate. She'd see money everywhere. Then, when the time was right, maybe months later, maybe when the girl confessed she had bills to pay or was behind on her rent, he would casually mention that he knew a way she could earn quick and easy money. He'd tell her how it worked. It was a victimless crime, he'd explain. They'd be "sticking it to the man."

None of the girls knew where Chris got his credit card data. When he referred to Max, it was as "the Whiz," an unnamable superhacker whom they'd never have the privilege of meeting. Chris's code name was "the

Dude." Now that his operation was purring, the Dude was paying the Whiz around $10,000 a month for the dumps—transferring the payments through a prepaid debit card called Green Dot.

Marketed to students and consumers with poor credit, a Green Dot Visa or MasterCard is a credit card without the credit: The consumer funds the card in advance with direct payroll deposits, transfers from a bank account, or cash. The last option made it an ideal money pipeline between Chris in Orange County and Max in San Francisco: Chris would drop in at a neighborhood 7-Eleven or Walgreens and purchase a Green Dot recharge number, called a MoneyPak, for any amount up to $500. He'd then IM or e-mail the number to Max, who'd apply it to one of his Green Dot cards at the company's website. He could then use the card for everyday purchases or make withdrawals from San Francisco ATMs.

Once his crew arrived, ready for work, Chris passed out their cards, separated into low-limit classic cards and high-limit gold and platinum. They should stick to small purchases for the classics, he'd remind them—$500 or so. With the high-limit plastic they should go for the big bucks, purchases from $1,000 to $10,000 dollars. The girls were all young, but affecting the privileged bearing of stylish Orange County youth they could walk into a Nordstrom's and snatch up a couple of $500 Coach bags without raising eyebrows, then cross to the other side of the mall and do the same thing at Bloomingdale's.

New cashers were always nervous at first, but once the first fake card was approved at the register, they were hooked. In no time they'd be sending Chris excited text messages from their shopping excursions: "Can we use amex at new bloomingdales?" or "I did over 7k on a mc! yeah!"

At the end of the day, they met Chris in a parking lot and transferred the purses trunk-to-trunk. He paid them on the spot, 30 percent of the retail value, and carefully recorded the transaction on a payout sheet like a real businessman. The handbags—elegant cloth and suede and gleaming buckles—would go in boxes until Chris's wife, Clara, could sell them on eBay.

As night fell over Villa Siena, the lights went on above the tennis courts and the outdoor fireplaces ignited. Miles away Chris and his crew were at a restaurant, ordering a celebratory dinner and a bottle of wine. As always, it was Chris's treat.

14

The Raid

"Nice TV!" said Tim, admiring the sixty-one-inch Sony plasma hanging on the wall. Charity, a compulsive reader, hated the new flat-screen, the way it dominated the living room in their new apartment, but Max loved his gadgets, and this one was more than a high-def toy. It was a symbol of the couple's newfound financial security.

Max's friends knew that he was into something, and not just because he was no longer struggling to make ends meet. Max had begun slipping Tim CD-ROMs burned with the latest exploits from the underground, giving the system administrator an edge in protecting his work machines. Then there were the odd comments at the monthly Hungry Programmers' dinner at Jing Jing in Palo Alto. When everyone was done describing their latest projects, Max would only offer a cryptic note of envy. "Wow, I wish I was doing something positive."

But nobody was pressing Max for the details of his new gig; they could only hope it was something quasilegitimate. The hacker scrupulously avoided burdening his friends with the knowledge of his double life, even as he slipped farther to the edge of their circle. Until the day one of his hacks followed him home.

. . .

It was 6:30 a.m. and still dark out when Chris Toshok awoke to the sound of his doorbell buzzing, the long continuous drone of someone holding their thumb on the button. Figuring it for a neighborhood drunk, he rolled over and tried to get back to sleep. Then the buzz broke into an insistent rhythm, *bzzz, bzzz, bzzz,* like a busy signal. He reluctantly crawled out of bed, grabbed his pants and a sweatshirt, and moved groggily down the stairs.

When he opened the door he found himself squinting into the glare of a flashlight.

"Are you Chris Toshok?" said a woman's voice.

"Uh, yes."

"Mr. Toshok, we're with the FBI. We have a warrant to search the premises."

The agent—a long-haired blonde—showed Toshok her badge and pressed a thin sheaf of papers into his hands. Another agent put a firm hand on his arm and guided him outside to the porch, clearing the doorway to admit a flood of suits into the house. They roused Toshok's roommate, then began tossing Chris's bedroom, riffling through his bookshelves and pawing through his underwear drawer.

The blonde, joined by a Secret Service agent, sat down with Toshok to explain why they were there. Four months earlier the source code for the unreleased first-person shooter Half-Life 2 had been stolen from the computers of Valve Software in Bellevue, Washington. It was swapped in IRC for a while and then showed up on file-sharing networks.

Half-Life 2 was perhaps the most anticipated game of all time, and the emergence of the secret source code had electrified the gaming world. Valve announced it would have to delay the launch of the game, and the company CEO issued a public call for Half-Life fans to help track down the thief. Based on sales of the original game, Valve valued the software at a quarter of a billion dollars.

The FBI had traced some of the hacking activity to Toshok's Internet IP address at his old house, the agent explained. The judge would go easier on Toshok if he told them where he'd stashed the source code.

Toshok protested his innocence, though he acknowledged that he knew about the breach. His old friend Max Vision was staying with him at the time of the intrusion, and he got very excited when the source code popped up online.

Hearing Max Vision's name sent the agents into double time—they nearly tripped over themselves to finish the search and get back to the office to prepare a warrant application for Max's new apartment. Chris watched gloomily as they gathered his nine computers, some music CDs, and his Xbox. The blonde agent registered the look on his face. "Yeah," she said, "this is going to be hard for you."

When Max heard about the raid, he knew he didn't have much time. He ran around his apartment stashing his gear. He hid an external hard drive in a stack of sweaters in the closet, another in a cereal box. One of his laptops fit under the sofa cushions; he hung a second one out the bathroom window in a garbage bag. Everything sensitive on his computers was encrypted, so even if they found his hardware, the agents wouldn't get any evidence of his hacking. But under the terms of his supervised release, he wasn't supposed to be using encryption at all. Moreover, it would be incredibly inconvenient to let the FBI take all of his computers.

The feds arrived in force, as many as twenty agents swarming like ants through the apartment. They found only the routine trappings of a San Francisco computer geek with hippie leanings: a bookshelf with Orwell's *1984,* Huxley's *Brave New World,* Orson Scott Card's sci-fi classic *Ender's Game,* and a smattering of Asimov and Carl Sagan. There was a bicycle, and stuffed penguins were strewn everywhere. Max loved penguins.

They discovered not one of Max's slapdash hiding spots, and this time, the hacker had nothing to say. The agents left without any evidence linking Max to the Valve intrusion, much less any hints of the crimes he was committing with Chris. Just a stack of CDs, a broken hard drive, and a vanilla Windows machine he'd left out as diversions.

But Charity had just learned what it meant to be in Max Vision's world. Max insisted he was innocent of the source code theft. It was probably the truth. There'd been several first-person shooter fans crawling around Valve's Swiss cheese network in anticipation of Half-Life 2. Max happened to be one of them.

The FBI later settled on a different Valve hacker: a twenty-year-old German hacker named Axel "Ago" Gembe, who admitted to his intrusions in e-mails to Valve's CEO, though he too denied stealing the code.

Gembe was already notorious for creating Agobot, a pioneering computer worm that did more than just spread from one Windows machine to another. When Agobot took over a machine, the user might not notice anything but a sudden sluggishness in performance. But deep in the PC's subconscious, it was joining a hacker's private army. The malware was programmed to automatically log in to a preselected IRC room, announce itself, and then linger to accept commands broadcast by its master in the chat channel.

Thousands of computers would report at once, forming a kind of hive mind called a botnet. With one line of text, a hacker could activate keystroke loggers on all the machines to capture passwords and credit card numbers. He could instruct the computers to open secret e-mail proxies to launder spam. Worst of all, he could direct all those PCs to simultaneously flood a targeted website with traffic—a distributed denial-of-service attack that could take down a top site for hours while network administrators blocked each IP address one at a time.

DDoS attacks started as a way for quarreling hackers to knock each other out of IRC. Then one day in February 2000, a fifteen-year-old Canadian named Michael "MafiaBoy" Calce experimentally programmed his botnet to hose down the highest-traffic websites he could find. CNN, Yahoo!, Amazon, eBay, Dell, and E-Trade all buckled under the deluge, leading to national headlines and an emergency meeting of security experts at the White House. Since then, DDoS attacks had grown to become one of the Internet's most monstrous problems.

Bots like Ago's marked the decade's major innovation in malware, inaugurating an era where any pissed-off script kiddie can take down part of the Web at will. Gembe's confession in the Valve hack provided the FBI with a golden opportunity to snare one of the innovators most responsible. The FBI tried to lure Gembe to America with an Invita-style job offer from Valve. After months of negotiations and telephone interviews with Valve executives, the hacker seemed ready to hop a flight to the States.

Then the German police intervened, arrested the hacker, and charged him locally as a youthful offender. Gembe was sentenced to one year of probation.

The raid on Max's house shook him, filling his head with unpleasant memories of the FBI's search warrant over the BIND attacks. Max decided he needed a safe house in the city, a place where he could ply his trade and store his data free from the threat of search warrants—something like Chris's Villa Siena plant.

Under an alias, Chris rented a second apartment for Max, a spacious penthouse in the Fillmore District, with a balcony and a fireplace—Max liked working by an open fire, and he'd joked that he could burn the evidence in an emergency.

Max tried to get home to Charity daily, but with a comfortable hacker safe house to retreat to, he began disappearing for days at a stretch, sometimes only emerging when his girlfriend interrupted his work with a prodding phone call.

"Dude, time to come home. I miss you."

As money started to flow into Max and Chris's joint operation, so did the mistrust. Some of the cashers in Chris's crew liked to party, and the constant presence of cocaine, ecstasy, and pot called to Chris like a forgotten melody. In February, he was pulled over near his home and arrested for driving under the influence. He began routinely vanishing with his comely employees for weekend-long bacchanals in Vegas: The day was for shopping; at night, Chris would snort some coke and take the girls out to the Hard Rock to party or snag a VIP table at the sleek Ghostbar atop the Palms, where he'd blow $1,000 on dinner and another grand on wine. Back in Orange County, he took a mistress—an eighteen-year-old woman he met through one of his cashers.

Max found both drugs and marital infidelity distasteful. But what really irked him was the financial arrangement. Chris was paying Max haphazardly—in whatever amount he felt like turning over at any given moment. Max wanted a straight 50 percent of Chris's profits. He was certain that Chris was making serious bank from their joint operation.

Chris tried to set him straight, and he e-mailed Max a detailed spreadsheet showing where the profits were going. Out of a hundred cards, maybe fifty worked, and only half of those could buy anything worth selling—the others were seeds and stems, cards with $500 security limits that were good only for trifles like gas and meals. Chris had expenses, too—spreading his hustle meant flying his crew to far-flung cities, and airline seats weren't getting any cheaper. Meanwhile, he was paying rent at Villa Siena for his credit card factory.

Max was unconvinced. "Call me back when you're not stoned."

The last straw came when Chris, three months after the Half-Life raid, suffered a close call himself. He'd driven up to San Francisco to meet with Max and make some carding runs at Peninsula malls. He and his crew were checked into adjacent rooms at the W, a posh hotel in the Soma district, when Chris got a call from the front desk. His credit card had been declined.

Hungover and fuzzy-headed from the flu, Chris took the elevator to the marbled lobby and pulled a new fake card from his swollen wallet. He watched as the clerk swiped it. It was declined. He produced another one, and it failed too. The third one worked, but by then the clerk was suspicious, and as the elevator was carrying Chris back to the twenty-seventh floor, she was picking up the phone and calling the credit card company.

The next knock on Chris's door was the San Francisco Police Department. They cuffed him and searched his rooms and car, seizing his Sony laptop, an MSR206, and his SUV, which had a fake VIN tag—Chris had experimented with renting cars using his plastic in Las Vegas, then sending them to Mexico to be fitted with clean VINs.

Chris was thrown in the county jail. His disappearance worried Max, but Chris bailed out quickly and confessed his blunder to his partner. Fortunately for him the police investigation went no further. Chris was sentenced a month later to three years of probation and ordered not to return to the W. He boasted afterward that he'd been a beneficiary of San Francisco's liberal justice system.

It was the kind of bullshit local bust that happened to Chris's girls all the time; that was why Chris kept a bail bondsman on retainer and even let him crash at his Villa Siena factory. But Max was furious. It was unforgivably sloppy for someone at Chris's level to be arrested carding a hotel room.

Max decided he could no longer rely exclusively on his partner. He needed a Plan B.

15

UBuyWeRush

The run-down strip mall was plunked down in that vast, flat interior of Los Angeles County. that doesn't make it onto postcards, far from the ocean and so distant from the hills that the squat stucco buildings could be a Hollywood set, the featureless sky behind them a blue screen to be filled in with mountains or trees in post-production.

Chris pulled his car into the trash-strewn parking lot. A marquee at the entrance gave top billing to the Cowboy Country Saloon, and below that it was the usual south Los Angeles mix: a liquor store, a pawnshop, a nail salon. And one more that was less usual: UBuyWeRush—the only retail sign in Los Angeles that was also a handle on CarderPlanet and Shadowcrew.

He walked into the front office, where an empty reception window suggested the sixty-cent-per-square-foot space had once been a medical clinic. On the wall a Mercator projection map of the world bristled with pushpins. Then Chris was greeted warmly by UBuy himself, Cesar Carrenza.

Cesar had come to the underground by a circuitous course. He graduated from the DeVry Institute in 2001 with a degree in computer programming, hoping to get an Internet job. When he couldn't find one, he decided to try his hand as an independent businessman on the Web.

From an ad in the *Daily Commerce,* he learned about an upcoming auction at a public storage facility in Long Beach, where the owners were selling off the contents of abandoned lockers. When he showed up he found the auction observed a very specific ritual. The manager, wielding an imposing bolt cutter, would snip off the defaulting renter's lock while the bidders watched, and then open the door. The bidders, about twenty of them, were expected to evaluate the contents from where they stood several feet away. The winner would then secure the unit with his own padlock and clear out the contents within twenty-four hours.

The experienced bidders were easy to spot: Padlocks hung from their belts, and they held flashlights to peer into the dark lockers. Cesar was less prepared but no less eager. He was the only bidder on the first lot, claiming a locker full of old clothes for $1.

He sold the clothes at a yard sale and on eBay for about $60. Figuring he'd found a nice little niche, Cesar started going to more auctions at storage facilities and business liquidations, breaking down large lots and moving them on eBay for a tidy profit. He put the money back into the business and opened his storefront in the Long Beach strip mall to accept consignments from neighbors with office furniture, lawn chairs, and unbranded jeans to sell online.

It was good, honest work—not like his last independent business. For most of the 1990s Cesar had been into credit card fraud. He was happier selling on eBay, but thinking about the past made him wonder if there was a market for the kind of gear he'd used as a crook. He ordered some MSR206s from the manufacturer and offered them for sale through the UBuyWeRush eBay store. He was impressed by how fast they were snapped up.

Then one of his new customers told him about a website where he could really sell. He introduced Cesar to Script, who approved UBuyWeRush as a CarderPlanet vendor. Cesar posted his introduction on August 8, 2003. "I decided to supply all you guys making the real big bucks," he wrote.

"So if you need me I sell card printers, card embossers, tippers, encoders, small readers and more. I know it sounds like advertising, but it's for you, a SAFE place to shop."

Business exploded overnight. Cesar built his own website, began vending on Shadowcrew, got an 800 number, and started accepting e-gold, an anonymous online currency favored by carders. He developed a reputation for excellent customer service. With customers in every time zone, he was scrupulous about answering the phone whenever it rang, day or night. It was always money on the other end of the line.

A canny businessman, he guaranteed same-day shipping and forged relationships with his rivals, so if he was caught short on an item, he could buy stock from a competitor to fill his orders and keep his customers happy. Strategic moves like that soon turned UBuyWeRush into the top supplier of hardware to a worldwide community of hackers and identity thieves. "Really good person, great to deal with," wrote a carder named Fear, advising a Shadowcrew newbie. "Don't scam UBuyWeRush cause he's a cool guy, and he'll keep your info on the downlow."

Cesar soon expanded his offerings to include hundreds of different products: skimmers, passport cameras, foil stampers, blank plastic, barcode printers, embossers, check paper, magnetic ink cartridges, even cable TV descramblers. Selling equipment wasn't in and of itself illegal, as long as he wasn't conspiring in its criminal applications. He even had some law-abiding customers who bought his gear to make corporate ID cards and school lunch vouchers.

Inundated with orders, Cesar ran a help-wanted ad in the classifieds and began hiring workers to inventory, pack, and ship his gear. As adjoining offices opened up, he annexed them for the extra storage space, doubling and then tripling his square footage. Fascinated by the global reach of his low-rent strip-mall operation, he bought a wall map, and every time he shipped to a new city he'd sink a pin into the location. After six months, the map was porcupined with pins throughout the United States,

Canada, Europe, Africa, and Asia. An impenetrable forest of metal grew southwest of Russia on the Black Sea. Ukraine.

Chris had become friends with Cesar. He'd even had him over for dinner, along with Mrs. UBuyWeRush, Clara, and Chris's two boys—well-mannered kids who stayed at the dinner table all the way through dessert. Chris particularly liked hanging out at Cesar's office. You never knew who would show up at UBuyWeRush. Carders too paranoid to have counterfeiting gear shipped even to a drop would make a pilgrimage to Los Angeles to pick up their items in person, opening the front door through their shirtsleeve to leave no prints and paying in cash. Foreign carders vacationing in California would stop by just to see the legendary warehouse with their own eyes and shake Cesar's hand.

On this day, the man walking in to pick up an MSR206 was the last person Chris expected to see in Cesar's shop, a six-foot-five hacker with a long ponytail.

Chris was stunned; Max rarely left San Francisco these days, and he hadn't said anything about coming to town. Max was equally surprised to see Chris. They exchanged pleasantries awkwardly.

There was only one reason Max would sneak into Los Angeles to buy his own magstripe encoder, Chris knew. Max had decided to stop sharing his most valuable data.

Max had become privy to one of the biggest security blunders in banking history, one that most consumers would never hear about, even as it enriched carders to the tune of millions of dollars.

The midsized Commerce Bank in Kansas City, Missouri, may have been the first to figure out what was going on. In 2003, the bank's security manager was alarmed to find that customer accounts were being sacked for $10,000 to $20,000 a day from cash machines in Italy—he would come in on a Monday and find his bank had lost $70,000 over the weekend. When he investigated, he learned that the victim customers had all fallen for a phishing attack aimed specifically at their debit card numbers and PINs.

But something didn't make sense: CVVs were supposed to prevent exactly this kind of scam. Without the CVV security code programmed onto the magnetic stripe of the real cards, the phished information shouldn't have worked at any ATM in the world.

He dug some more and discovered the truth: His bank simply wasn't checking the CVV codes on ATM withdrawals, nor on debit card purchases, where the consumer enters the PIN at the register. In fact, the bank couldn't perform such a check consistently if it wanted to; the third-party processing network used by the bank didn't even forward the secret code. The Italian phishers could program any random garbage into the CVV field, and the card would be accepted as the real thing.

The manager moved the bank to another processing network and reprogrammed his servers to verify the CVV. The mysterious withdrawals from Italy halted overnight.

But Commerce Bank was just the beginning. In 2004, nearly half America's banks, S&Ls, and credit unions still weren't bothering to verify the CVV on ATM and debit transactions, which is why America's inboxes were being flooded with phishing e-mails targeting PIN codes for what the carders called "cashable" banks.

Citibank, the nation's largest consumer bank by holdings, was the most high-profile victim. "This e-mail was sent by the Citibank server to verify your e-mail address," read a message spammed from Russia in a September 2003 campaign. "You must complete this process by clicking on the link below and entering in the small window your Citibank ATM/Debit Card number and PIN that you use on ATM."

A more artful message in 2004 capitalized on consumers' well-founded fears of cybercrime. "Recently there have been a large number of identity theft attempts targeting Citibank customers," read the spam, emblazoned with Citi's iconography. "In order to safeguard your account, we require that you update your Citibank ATM/Debit card PIN." Clicking on the link took customers to a perfect simulacrum of a Citibank site, hosted in China, where the victim would be prompted for the data.

Good for direct cash, PINs were the holy grail of carding. And it was CarderPlanet's King Arthur who was most successful in the quest. King, as he was known to his friends, ran an international ring that specialized in hitting Citibank customers, and he was a legend in the carding world. One of King Arthur's lieutenants, an American expat in England, once let it slip to a colleague that King was making $1 million a week from the global operation. And he was just one of many Eastern Europeans running cash-outs in America.

Max plugged himself into the Citibank cash-outs in his own way: He Trojaned an American mule named Tux, and started intercepting the PINs and account numbers the carder was getting from his supplier. After a while, he contacted the source—an anonymous Eastern European whom Max suspected of being King Arthur himself—and told him candidly what he'd done: Tux, he said, had been guilty of the crime of slipshod security. For good measure, Max claimed falsely that the mule had been ripping off the supplier.

The supplier cut off Tux on the spot and began providing Max with his PINs directly, anointing the hacker as his newest cash-out mule.

When the PINs first started rolling in, Max had passed them all to Chris, who tore into them with a vengeance. Chris would pull $2,000 in cash—the daily ATM withdrawal limit—and then send his girls out to make in-store debit purchases with the PINs until the account was drained dry. He was raping the cards. Max didn't like it. The whole point of a cash-out was to get *cash,* not merchandise that sold for a fraction of its worth. With a little finesse, the PINs could be producing a lot more liquid.

Then it occurred to him he didn't need his partner at all for this particular operation.

When he returned from UBuyWeRush with his very own MSR206, Max went into business for himself. He programmed a stack of Visa gift cards with the account data and wrote each card's PIN on a sticky note affixed to the plastic. Then he'd get on his bicycle or take a long meandering walk through the city, visiting small, customer-owned cash machines at locations free of surveillance cameras.

He'd enter the PIN, then the withdrawal amount, and *chump, chump, chump, chump,* the ATM spat out cash like a slot machine. Max would pocket the money, write the new, lower account balance on the Post-it, then look around discreetly to make sure he hadn't drawn any attention before drawing the next card from his deck. To keep his prints off the machines, he'd press the buttons through a piece of paper or with his fingernails, or coat the pads of his fingers with hydroxyquinoline—a clear, tacky antiseptic sold in drugstores as the liquid bandage New-Skin.

Max dutifully sent a fixed percentage of his take to Russia via Western Union MoneyGram, per his agreement with the supplier. He was an honest criminal now, doing straightforward business in the underground. And even after getting his own magstripe writer, Max continued to give some of his PINs to Chris, who continued tapping his crew to burn through the cards aggressively.

On the surface, Max's ATM visits weren't much of a Robin Hood operation, but Max took moral solace in the fact that the cash-outs always ended with the cards being canceled. That meant the fraudulent withdrawals were being discovered, and Citibank would be forced to reimburse its customers for the thefts.

After some months, Max built a nice nest egg from Citibank's losses: He moved with Charity to a $6,000-a-month house rental in San Francisco's Cole Valley and installed a safe for his profits: $250,000 in cash.

His earnings were just a tiny piece of the losses from the CVV gaffe. In May 2005, a Gartner analyst organized a survey of five thousand online consumers and, extrapolating the results, estimated that it had cost U.S. financial institutions $2.75 billion. In just one year.

16

Operation Firewall

There was something fishy going on with Shadowcrew.

Max kept his presence on the Internet's top crime site low-key; to him, Shadowcrew was just a hunting ground conveniently stocked with hackable carders. But in May 2004, a Shadowcrew administrator made an offer on the board that got Max's attention. The admin, Cumbajohnny, was announcing a new VPN service just for Shadowcrew members.

A VPN—virtual private network—is typically used to provide telecommuters with access to their employer's network from home. But a trustworthy underground VPN appealed to carders for another reason. It meant every byte of traffic from their computers could be encrypted—immune to sniffing by a nosy ISP or a law enforcement agency with a surveillance warrant. And any attempt to trace their activities would get no farther than Cumbajohnny's own data center.

Cumbajohnny was a recent addition to Shadowcrew's leadership—a former moderator who was growing in power and influence and changing the mood on the board. Some other admins were complaining about a new mean-spiritedness on the forum. Banner ads appeared at the top of the site: "Stop talking. Do Business. Advertise here. Contact Cumbajohnny." Shadowcrew was taking on the feel of the Las Vegas strip, with flashy ads promising a lifestyle of partying, beautiful women, and piles and piles of cash.

Gollumfun, an influential founder, had already publicly retired from the site when another founder named BlackOps announced he was leaving as well. "Shadowcrew has been reduced from its once lustrous form to a degrading environment of children who lack knowledge, the skills or desire to interact with other members in a positive way," he wrote. "Gone are the well thought out tutorials; gone are the well-respected members; and gone is the civility. No longer do we help the newbies find their way, we simply flame them to death until they leave and then complain that there aren't any new members."

"BlackOps, you will be missed, thank you for your services," Cumbajohnny wrote tactfully. "SC is changing, and for the best."

Max paid little attention to the politics of the carding scene. But the VPN announcement made him uneasy. It turned out Cumbajohnny had been privately selling his VPN service to Shadowcrew's leaders for three months. Now, Cumbajohnny wrote, any Shadowcrew member in good standing could buy the same peace of mind for $30 to $50 a month.

But VPNs have one well-known weakness: everything transpiring over the network has to be funneled through a central point, unencrypted and vulnerable to eavesdropping. "If the FBI, or whoever, really wanted to they could get into the datacenter and change some of the configs on the VPN box and start logging, and then you would be kinda screwed," one member noted. "But that is just straight paranoia," he conceded.

Cumbajohnny reassured him. "No one can touch the VPN without me knowing about it."

Max wasn't convinced. In his white-hat days, he'd written a program for the Honeynet Project called Privmsg—a PERL script that took the data from a packet sniffer and used it to reconstruct IRC chats. When an intruder was lured into cracking one of the project's honeypots, the attacker would often use the system to hold online conversations with his fellow hackers. With Privmsg, the white hats could see the whole thing. It had been a strong innovation in hacker tracking, turning passive honeypots into digital wiretaps and opening a window into the underground's culture and motives.

Max could see the same wiretap tactic at play now in Cumbajohnny's VPN offer. There was other evidence, too; while hacking random carders, he saw a message to a Shadowcrew administrative account that read like a federal agent giving orders to an informant. Max couldn't shake the feeling that someone was turning Shadowcrew into the ultimate honeypot.

After talking it over with Chris, Max posted several messages to Shadowcrew summarizing his doubts. The posts disappeared at once.

Max's suspicions were right on the money.

The NYPD had nabbed Albert "Cumbajohnny" Gonzalez nine months earlier pulling cash out of a Chase ATM on New York's Upper West Side. Originally from Miami, Gonzalez was twenty-one years old and the son of two Cuban immigrants. He was also a longtime hacker who'd been dedicated enough to trek to Vegas for the 2001 Def Con.

The Secret Service interviewed Gonzalez in custody and quickly ascertained his worth. The hacker was living in a $700-a-month garden apartment in Kearny, New Jersey, had $12,000 in credit card debt, and was officially unemployed. But as "Cumbajohnny," he was a trusted confidant and colleague of carders around the world and, most importantly, a moderator at Shadowcrew.

He was in the belly of the beast, and properly handled, he might strike a deathblow against the forum.

The Secret Service took over the case and sprang Gonzalez to use him as an informant. The VPN was the agency's masterstroke. The equipment was bought and paid for by the feds, and they'd obtained wiretap warrants for all the users. Cumbajohnny's carder-only VPN service was an invitation to an Internet panopticon.

Shadowcrew's biggest players were drawn inexorably into the Secret

Service's surveillance net. The tapped VPN laid bare all the wheeling and dealing the carders kept off the public website—the hard negotiating that unfolded mostly in e-mail and over IM.

There were deals every day and every night, with a weekly surge in trading Sunday evenings. The transactions ranged from the petty to the gargantuan. On May 19, agents watched Scarface transfer 115,695 credit card numbers to another member; in July, APK moved a counterfeit UK passport; in August, Mintfloss sold a fake New York driver's license, an Empire Blue Cross health insurance card, and a City University of New York student ID card to a member in need of a full identification portfolio. A few days later, another sale by Scarface, just two cards this time; then MALpadre bought nine. In September, Deck sold off eighteen million hacked e-mail accounts with user names, passwords, and dates of birth.

The Secret Service had fifteen full-time agents combing through the activity—every purchase would be another "underlying offense" in a grand jury indictment. And the best part was, many of Shadowcrew's denizens were unwittingly paying the Secret Service for the privilege of being monitored.

But running a game against hackers was never cut-and-dried, as the agency learned on July 28, 2004. That was when Gonzalez informed his handlers that a carder named Myth, one of King Arthur's cashers, had somehow obtained one of the agency's confidential documents about Operation Firewall. Myth had been boasting about it in an IRC chat room.

The feds told Gonzalez to find the source of the leak, and fast. As Cumbajohnny, Gonzalez made contact with Myth and learned that the documents represented just a few droplets in a full-blown Secret Service data spill. Myth knew about subpoenas issued in the Shadowcrew probe and had even discovered that the agency was monitoring his own ICQ account. Fortunately, the documents didn't mention an informant.

Myth refused to tell Gonzalez who his source was but agreed to arrange an introduction. The next day, Gonzalez, Myth, and a mystery hacker using the temporary handle "Anonyman" met on IRC. Gonzalez worked to gain Anonyman's trust, and the hacker finally revealed himself as Ethics, a vendor whom Cumba already knew on Shadowcrew.

The leak was starting to make sense. In March, the Secret Service had noticed Ethics was selling access to the database of a major wireless carrier, T-Mobile. "I am offering reverse lookup of information for a T-Mobile cell phone, by phone number," he wrote in a post. "At the very least, you get name, SSN, and DOB. At the upper end of the information returned, you get Web username/password, voicemail password, secret question/answer."

T-Mobile had failed to patch a critical security hole in a commercial server application it had purchased from the San Jose, California, company BEA Systems. The hole, discovered by outside researchers, was painfully simple to exploit: An undocumented function allowed anyone to remotely read or replace any file on a system by feeding it a specially crafted Web request. BEA produced a patch for the bug in March 2003 and issued a public advisory rating it a high-severity vulnerability. In July of that year, the researchers who discovered the hole gave it more attention by presenting it at the Black Hat Briefings convention in Las Vegas, an annual pre–Def Con gathering attended by 1,700 security professionals and corporate executives.

Ethics learned of the BEA hole from the advisory, crafted his own twenty-line exploit in Visual Basic, then began scanning the Internet for potential targets who had failed to patch. By October 2003, he hit pay dirt at T-Mobile. He wrote his own front end to the customer database to which he could return at his convenience.

At first, he used his access to raid the files of Hollywood stars, circulating grainy candid photos of Paris Hilton, Demi Moore, Ashton Kutcher, and Nicole Richie stolen from their Sidekick PDAs. It was evident now that he'd gotten into a Secret Service agent's Sidekick as well.

A simple Google search on Ethics's ICQ number turned up his real name on a 2001 résumé seeking computer security work. He was Nicholas Jacobsen, a twenty-one-year-old Oregonian who'd recently relocated to Irvine, California, to take a job as a network administrator. All that was left was to confirm which Secret Service agent was violating policy by accessing sensitive material on his PDA.

That's where Gonzalez proved his worth again. Now that he was buddies with Cumbajohnny, Ethics hit up the Shadowcrew leader for an account on his much-touted VPN, figuring it would be a safer way to access T-Mobile.

Gonzalez happily obliged, and his Secret Service handlers got to watch as Ethics surfed to T-Mobile's customer service website and logged in with the user name and password of New York agent Peter Cavicchia III, a veteran cybercrime officer who'd distinguished himself by busting a former AOL employee for stealing ninety-two million customer e-mail addresses to sell to spammers.

The leak had been found. Cavicchia would quietly retire a few months later, and Ethics was added to the list of Operation Firewall targets.

There was just one more threat to the investigation, and, bizarrely, it was coming from one of the FBI's underground assets.

David Thomas was a lifelong scammer who'd discovered the crime forums in the Counterfeit Library days and soon became addicted to the high-speed deal making and criminal camaraderie. Now forty-four years old, El Mariachi, as he styled himself, was one of the most respected members in the carding community, assuming the role of mentor to younger scammers and dispensing advice on everything from identity theft to basic life lessons gleaned from decades on the fringe.

His experience, though, didn't immunize him from the hazards of his profession. In October 2002, Thomas showed up in an office park in Issaquah, Washington, where he and his partner had rented a drop for one of CarderPlanet's founders. They were hoping to claim $30,000 in Outpost .com merchandise ordered by the Ukrainian. Instead, they found local police waiting for them.

The police arrested Thomas, and a detective read him his Miranda rights and gave him a form to sign acknowledging he understood them. Thomas scoffed at the idea of a local cop trying to question him. "You don't know who you have here," he said. He urged the detective to call in the feds; the Secret Service would know who El Mariachi was, and he could give them a case involving Russians and "millions of dollars."

A Secret Service agent visited him in the county jail but wasn't impressed by Thomas's $30,000-drop business. Then an FBI agent from the Seattle field office showed up. On the second meeting, the agent brought along an assistant U.S. attorney and an offer: The feds couldn't help Thomas with his local case, but when he got out he could go to work for the Northwest Cyber Crime Task Force in Seattle.

It would be an intelligence-gathering mission, an official designation for an FBI operation with no predetermined targets. The bureau would get Thomas a new computer, put him up in a nice apartment, pay all of his expenses, and give him $1,000 a month in spending money. In return, Thomas would gather information on the underground and report it back to the task force.

Thomas hated snitches, but he liked the idea of being paid to observe and comment on the underground with which he'd become obsessed. Intelligence gathering wasn't the same as snitching, he reasoned, and he could use the material he collected to write a book about the carding scene, something he'd been thinking a lot about lately.

He also knew exactly how to gather the information the task force was after.

Thomas was released from jail five months after his arrest. And in April, the FBI gained a new asset in the war on cybercrime: El Mariachi and his brand-new government-funded crime forum, the Grifters.

From his bureau-rented corporate apartment in Seattle, El Mariachi was soon gathering information on his fellow carders, particularly the Eastern Europeans. But though Thomas was working for the FBI, he didn't exactly feel kinship with other government assets, and the VPN announcement convinced him—correctly—that Cumbajohnny was a federal informant.

Thomas became fixated on exposing his rival. Ignoring admonishments from his FBI handler, he continuously called out Gonzalez on the forums. Gonzalez, too, seemed to have it in for El Mariachi—he dug up a copy of the police report from Thomas's Seattle arrest and circulated it among the Eastern European carders, drawing their attention to the part where Thomas offered to help catch Russians. A full-blown proxy war had broken out between the FBI and Secret Service, by way of two informants.

It was a bad time to be distracting the Eastern Europeans with American carder drama. In May 2004, one of CarderPlanet's Ukrainian founders was extradited to the United States, after being arrested on vacation in Thailand. The next month, the British national police moved in on the site's only native English-speaking administrator in Leeds.

Script, getting heat from the Orange County FBI and the U.S. Postal Inspection Service, had already retired from the site, leaving King Arthur in charge. On July 28, 2004, King made an announcement.

"It is time to tell you the bad news—the forum should be closed," he wrote. "Yes, it really means closed and there are a lot of reasons for that."

In broken English he explained that CarderPlanet had become a magnet for law enforcement agencies around the world. When carders were busted, police interrogators badgered them with questions about the forum and its leaders. Under the relentless pressure, he implied, even he might slip up. "All of us are just people and all of us can make mistakes."

By closing CarderPlanet, he would be depriving his enemies of their greatest asset. "Our forum held them well informed and up to date, and on our forum they and the bank employees just have been raising their level of proficiency and knowledge," he wrote.

"Now all of thing will be the same but they will not know where the wind blows from and what to do."

With that farewell note, King Arthur, almost certainly a millionaire ten times over, became a carder legend. He would be remembered as the one who gently folded the great CarderPlanet before anyone else could enjoy the pleasure of taking it down.

Shadowcrew's leaders wouldn't be so lucky. In September, the FBI pulled the plug on Thomas's operation and gave him a month to move out of his apartment—ending his war with Cumbajohnny. The next month, on October 26, sixteen Secret Service agents gathered in a Washington command center to drop the hammer on Operation Firewall. Their targets were marked on a map of the United States filling a wall of computer displays. Every one of them would be at home, the agents knew; at the Secret Service's behest, Gonzalez had called an online meeting for that evening, and nobody said no to Cumbajohnny.

At nine p.m., agents armed with MP5 semiautomatic assault rifles burst into Shadowcrew members' homes around the country, grabbing three founders, T-Mobile hacker Ethics, and seventeen other buyers and sellers. It was the biggest crackdown on identity thieves in American history. Two days later, a federal grand jury handed down a sixty-two-count conspiracy indictment and the Justice Department went public with Operation Firewall.

"This indictment strikes at the heart of an organization that is alleged to have served as a one-stop marketplace for identity theft," Attorney General John Ashcroft boasted in a press release. "The Department of Justice is committed to taking on those who deal in identity theft or fraud, whether they act online or off."

With Gonzalez's help, the Secret Service locked Shadowcrew's remaining four thousand users out of the site and swapped in a new front page featuring a Secret Service banner and an image of a prison cell. The new page struck the Shadowcrew tagline, "For Those Who Like to Play in the Shadows," and substituted a new motto: "You Are No Longer Anonymous!!"

Panicked carders around the around the world soaked up the news reports and watched the television coverage, worrying for themselves and their fallen compatriots. They collected on a small forum called Stealth Division to assess the damage and take a head count of survivors. "I am scared to death for my family right now—for my children," wrote one cyberthief. "I just learned that my every move has been recorded."

Slowly, they realized that Cumbajohnny wasn't on the list of defendants. That's when he logged in to make a final appearance.

"I want everyone to know I'm on the run and I had no fucking idea the USSS had the capabilities of doing what they did," Gonzalez wrote. "From the news articles I can tell they've wiretapped my VPN and wiretapped the Shadowcrew server. This is my last post, good luck everyone."

Nick Jacobsen, Ethics, was kept out of the press release and quietly indicted separately in Los Angeles—his intrusion into the Secret Service's e-mail wouldn't emerge until well after the agency had collected its accolades for Operation Firewall. Even then, the dragnet was a clear victory for the government. CarderPlanet was shuttered, and now Shadowcrew was closed for good, and its leaders—save Gonzalez—were in jail.

The carders were confused, paranoid, and, for the moment, homeless. "It will take years and years for any message board like Shadowcrew to build up," wrote one. "And when or if it does, law enforcement will bust it again.

"And knowing what can be done, I doubt anyone will take the risk of putting another one up."

17

Pizza and Plastic

On the top floor of the Post Street Towers, Max's computers sat on the wood-veneer floor, silent and cool. Outside the bay window, shops and apartments were ready to unwittingly feed him bandwidth through his oversized antenna.

Max had gone dormant for a few months after accumulating a pile of cash from the Citibank operation; he'd abandoned his penthouse apartment and put his hacking on the back burner. But he couldn't stay away long. He'd asked Chris to rent him a new safe house, one with more neighborhood Wi-Fi options than the last. "I just need a closet, I don't need any space," he'd said.

Chris had delivered. There was ample Wi-Fi swimming around the Post Street Towers, and the apartment was indeed a closet: a three-hundred-square-foot studio that seemed scarcely larger than a prison cell. Decked out in blond wood, with a Formica counter, a full-sized fridge, and a bed that unfolded from the wall, it was a clean and functional McApartment, bare of all distractions and able to provide the necessities for Max's all-night hacking sprees. The high turnover in the building made him anonymous. Chris just had to flash a fake ID at the rental office, pay a $500 deposit, and sign the six-month lease.

Once his computers were plugged in and his antenna was latched on to some patsy's network, Max wasted little time in getting back on the job.

As ever, he targeted fraudsters, and he developed some novel ways to steal from them. He monitored the alerts put out by an organization called the Anti-Phishing Working Group, staying on top of the latest phishing attacks. The alerts included the Web addresses of the phishing sites linked to the forged e-mails, allowing Max to hack the phishers' servers, resteal the stolen data, and erase the original copy, frustrating the phishers and grabbing valuable information at the same time.

Other attacks were less focused. Max was still plugged into the white-hat scene, and he was on the private mailing lists where security holes often appeared for the first time. He had machines scanning the Internet day and night for servers running vulnerable software, just to see what he'd turn up. He was scanning for a Windows server-side buffer overflow when he made the discovery that would lead to his public entry into the carding scene.

His scanning put him inside a Windows machine that, on closer inspection, was in the back office of a Pizza Schmizza restaurant in Vancouver, Washington; he knew the place, it was near his mother's house. As he looked around the computer, he realized the PC was acting as the back-end system for the point-of-sale terminals at the restaurant—it collected the day's credit card transactions and sent them in a single batch every night to the credit card processor. Max found that day's batch stored as a plain text file, with the full magstripe of every customer card recorded inside.

Even better, the system was still storing all the previous batch files, dating back to when the pizza parlor had installed the system about three years earlier. It was some fifty thousand transactions, just sitting there, waiting for him.

Max copied the files, then deleted them—they weren't needed by Pizza Schmizza; in fact, just storing them in the first place was a violation of Visa's security standards. After sorting and filtering out the duplicate and expired cards, he was left with about two thousand dumps.

For the first time, Max had a primary source, and they were virgin cards, almost guaranteed to be good.

Chris had been complaining about the staleness of some of Max's dumps. That would end now. A customer could walk into the Pizza Schmizza and order a twelve-inch pie for his family, and his credit card could be on Max's hard drive while the leftovers were still cooling in the garbage. Once he was done organizing his numbers, Max gave Chris a taste. "These are extremely fresh," he said. "They're from two days ago."

There was no way that Chris and his crew could metabolize the fifty dumps a day coming from the Pizza Schmizza. So Max decided to make his first forays into vending in the carding scene.

Chris offered to handle the sales in exchange for half the profits. Chris's recklessness still concerned Max—Chris had nearly been arrested buying gold in, of all places, India, fleeing the country one step ahead of the police. But Chris knew too much about Max for the hacker to just cut him loose, so he agreed to let Chris act as his representative to the underground. Chris soon claimed success in marketing Max's dumps, until Max—who had a back door on Chris's computer—figured out that Chris was actually using the magstripe data himself, getting a 50 percent price break by claiming to have resold them. Economically, it was all the same. But Max couldn't help feeling cheated yet again.

Max turned to someone who might be easier to control: a teenage carder from Long Island named John Giannone who had become Chris's sidekick.

Giannone was a smart middle-class kid with a coke habit and burning desire to be a ruthless, badass cyberpunk. His early ops failed to impress: He boasted to another carder that he'd once pushed all the buttons on an elevator before getting off, so the next passenger would have to stop at every floor. On another occasion, he claimed, he walked into a bank and wrote a note on the back of a deposit slip: "This is a robbery. I have a bomb. Give me money or I'll blow the bank." Then he put the slip back on the pile as a surprise for the next customer.

When he was seventeen, Giannone joined Shadowcrew and CarderPlanet under the handle MarkRich, and started participating in small operations. His reputation went south when he was busted carding plane tickets and a rumor spread that he'd snitched on a forum regular while in juvenile hall.

Undaunted, Giannone paid a more established carder for the exclusive right to take over his handle and reputation. As "Enhance," the teen became more bold but not more successful. In May 2003, copying an extortion tactic perfected by the Russians, he borrowed a hacker's botnet and launched a DDoS attack against JetBlue, taking down the airline's website for some twenty-five minutes before sending an e-mail demanding $500,000 in protection money. But JetBlue paid him neither cash nor the respect a cybergangster deserved. "We will forward this to the appropriate law enforcement agencies," the company wrote. "Yesterday's outage was due to a system upgrade."

When Max found Giannone with his Free Amex hack, the teen was running his operations from the computer in his mother's bedroom. But Max and Chris had looked over Giannone's files and decided he could be partner material. Chris in particular may have seen something of himself in the young, coke-snorting gangster wannabe. Giannone was already a regular visitor to Orange County—he liked vacationing in the sun—and the two began partying together. Chris called his apprentice "the Kid."

Max knew everything about Giannone, while Giannone knew virtually nothing about him. For Max, it was an ideal arrangement for a partnership. Giannone made some sales of Max's dumps and then introduced Max to other carders interested in making buys over ICQ. Max set up a new online identity for his vending: "Generous."

Dealing with strangers was a big step for Max, and he took elaborate precautions to stay safe. When using carder forums or instant-messaging services, he'd bounce his connection through his private network of hacked PCs around the world—ensuring nobody could easily trace him even as far as his hacked WiFi. He disguised his writing style online for fear that some ill-considered turn of phrase or choice of punctuation might be matched to one of Max Vision's security white papers or Bugtraq posts—the FBI had once remarked on the copious ellipses in his anonymous note to Lawrence Berkeley Laboratory during the BIND attacks.

To collect revenue, he accepted payment through an anonymous e-gold account linked to an ATM card. Giannone helped him with a second remittance system. The teenager established a business account at Bank of America for a car repair shop called A&W Auto Clinic, then sent Max the magstripe data and PIN code for his ATM card, allowing Max to clone the card with his MSR206. Dumps buyers in the United States could make a cash deposit for A&W at their nearest Bank of America branch, which Max could then withdraw at his leisure with his cloned ATM card.

Max didn't need the money the way he used to. He'd squandered most of his nest egg from the Citibank cash-outs, frittering it away on everything from handouts for the homeless to a $1,500 Sony AIBO robotic dog. But he wasn't broke yet, and Charity had just started a well-paying job as a system administrator at Linden Lab, the brick-and-mortar home of Second Life—a fully realized three-dimensional online universe growing by thousands of inhabitants a month.

There was just one reason he was upping the ante now. He'd become addicted to life as a professional hacker. He loved the cat-and-mouse games, the freedom, the secret power. Cloaked in the anonymity of his safe house, he could indulge any impulse, explore every forbidden corridor of the Net, satisfy every fleeting interest—all without fear of consequence, fettered only by the limits of his conscience. At bottom, the master criminal was still the kid who couldn't resist slipping into his high school in the middle of the night and leaving his mark.

18

The Briefing

In a briefing room near Washington, two dozen male faces filled a computer monitor on the wall, some scowling for a mugshot, others smiling for a passport photo. A couple of them looked like teenagers barely out of puberty; others were older, unkempt and vaguely dangerous in appearance.

Around the table a handful of FBI agents in suits and ties stared back at the faces of the international computer underground. For one of the agents, a lot of things were suddenly making sense.

At thirty-five years old, J. Keith Mularski had been an FBI agent for seven years. But he'd been on the computer crime beat for just four months, and he had a lot to learn. Enthusiastically friendly and quick to laugh, Mularski had wanted to be an FBI agent since his freshman year at Pennsylvania's Westminster College, when a bureau recruiter came in to speak to one of his classes. He'd held on to the list of qualifications even as he walked a more pedestrian career path, starting as a furniture salesman in Pittsburgh, then working his way up to a position as operations manager for a national furniture chain with fifty employees reporting to him at four stores.

In 1997, after eight years of waiting, he finally decided he was ready for the FBI. After a yearlong application process and sixteen weeks of training at the FBI academy in Quantico, he was sworn in as an agent in July 1998.

As part of the bureau's graduation ritual, the newly minted agent was instructed to rank all the FBI field offices in order of assignment preference. He rated his hometown of Pittsburgh as number one—it was where Mularski had grown up, gone to school, and met his wife. His chances of transferring there evaporated the next month, when Islamic terrorists bombed U.S. embassy buildings in Kenya and Tanzania. Veteran FBI agents were dispatched from the Washington, DC, field office to investigate the attacks, and Mularski was one of fifteen fresh recruits sent to fill the vacancies in DC—the city marked thirty-second on his list.

Almost overnight Mularski went from managing furniture stores to working on some of the FBI's most important, and highly classified, investigations. When, in 1999, a listening device was found in an office on the top floor of the State Department's headquarters, he was part of the team that identified a Russian diplomat monitoring the transmitter from outside. In 2001, he helped bring down Robert Hanssen, a fellow counterespionage agent who'd been secretly spying for the KGB and its successor agency for twenty years.

It was heady work, but the secrecy chafed Mularski: He held a top-secret clearance and couldn't talk about his job with outsiders—even his wife. So when headquarters announced openings for two experienced agents to kick-start an ambitious cybercrime initiative in Pittsburgh, he saw a chance to go home and step out of the shadows at the same time.

His new job wouldn't be in an FBI office. He was assigned to the civilian office of an industry nonprofit group in Pittsburgh called the National Cyber Forensics and Training Alliance. The NCFTA had been formed by banks and Internet companies a couple of years earlier to track and analyze the latest scams targeting consumers online—mostly phishing attacks. Mularski's job wouldn't consist of chasing individual scams—in isolation, each round of phishing was too small to meet the FBI's minimum loss threshold of $100,000. Rather, he would be looking for trends that pointed to a common culprit—a group or a single hacker—responsible for a large number of cyberthefts. Then he'd shop the results to the various FBI field offices and, hopefully, hand off the investigation.

It was passive intelligence gathering, meticulous but unexciting. Mularski wasn't in charge of the cases, and he never got the satisfaction of putting handcuffs on a bad guy. But for the first time in seven years, he could talk about his work with his wife over dinner.

Now he was back in the DC area for his first briefing on the carding scene. At the head of the room was Postal Inspector Greg Crabb, a solidly built man with world-weary eyes who worked in the post office's international fraud unit. Crabb had stumbled upon the carding underground in 2002 while tracking a software counterfeiter with a sideline in credit card fraud. Since then, he'd been on the ground in twenty-five countries, working with local police to make busts and building a massive database of raw intelligence on the growing community: nicknames, IP addresses, instant messages, and e-mails of more than two thousand people. He'd become the government's top expert on the scene, but the enormity of his crusade now threatened to overwhelm him. So he'd come to the FBI for help.

The briefing for about half a dozen FBI agents was held at a nondescript Calverton, Baltimore, office where the bureau ran its Innocent Images anti–child porn operation. Speaking slowly in a rumbling, midwestern twang, the postal inspector weighed each word like a parcel as he ran through the history of the scene: CardersLibrary spawning CarderPlanet, the legend of King Arthur, the influence of the Russians and Ukrainians, and the rise and fall of Shadowcrew. He threw up a screenshot of CarderPlanet to show the underground's structure: A site operator was the don. Admins were capos. It was a metaphor to which the FBI was institutionally attuned; hackers were the new mafia.

Operation Firewall, Crabb explained, had left the carders scattered, paranoid, and disorganized. But they were rebuilding. And unlike before, with Shadowcrew, there was no singular target to go after. Instead, a slew of new, smaller forums was popping up. Crabb didn't say it, but the Secret Service had treated the carders with half a dose of penicillin; the survivors were immune and plentiful.

Mularski hung on every word. In his brief time at the NCFTA, the agent had seen patterns in the raw intelligence bubbling up from the underground: references to nicknames, coded messages, and forums. It made sense now. It was the carders organizing themselves again.

When Crabb wrapped up his talk and the other agents began to file out, Mularski approached the postal inspector at the head of the table and extended his hand enthusiastically. "This stuff is fascinating," he said. "I'd love to work with you. I'd love to partner up with you."

Crabb was surprised by the suggestion; in his experience, a more typical proposal from an FBI agent might take the form "Give me all your information. Thanks, bye." He met with Mularski and his boss privately and gave the agents a more thorough rundown on the carder scene.

Mularski returned to Pittsburgh, his head swimming. He'd thought he'd left behind the world of Russian spies, double agents, and secret identities. He'd been wrong. And the safe, satisfying routine of his new job was about to be shattered.

19

Carders Market

Try as he might, Max couldn't get situated on any of the new forums sprouting in Shadowcrew's ruins. They were all corrupt, run by dumps vendors hostile to outside competition. In a way, it was a blessing. He could never really trust any of the sites; he knew all too well that the scene was rank with cops and informants.

He finally made up his mind that if he was going to vend, the only sensible venue would be a site he personally controlled. Still thinking of himself as Robin Hood, he came up with the perfect name for his own forum: Sherwood Forest.

Chris approved of the plan—he liked the idea of vending his counterfeit credit cards and driver's licenses in a safe environment—but hated the name. As an exercise in branding, "Sherwood Forest" wasn't going to cut it for a criminal marketplace. The partners went back to the drawing board, and in June 2005 Max used a fake name and bogus address in Anaheim to register Cardersmarket.com.

It was a critical time for Max: He was near the end of his federal supervised release, and if he could make it until midnight, October 10, 2005, he would be a free agent, no longer obliged to play the role of an underemployed computer consultant for the benefit of his probation officer. It should have been easy enough to survive a few more months.

Besides Chris, there were only two people who knew about Max's double life, both Chris's friends: Jeff Norminton and Werner Janer, the real estate fraudster who wrote Charity a $5,000 check that helped bootstrap Max's hacking operation.

Then, in September 2005, Werner Janer got busted.

Since hooking up with Max, Chris had been dropping Janer a few cards here and there—maybe eighty over three years—in exchange for 10 percent of whatever Janer netted from his in-store purchases. That month Janer asked for another batch of two dozen cards—a money shortage had forced him to sell the family home in Los Angeles, and he'd moved to Westport, Connecticut, to make a new start of it. Soon after his arrival he was robbed by a criminal associate of nearly all the proceeds of the house sale, and he needed an income boost to support himself and his wife and three children.

When Chris's FedEx arrived, Janer, an avid watch collector, headed straight to Richard's of Greenwich, a men's clothing and accessory store that kept an inventory of high-end timepieces. Janer had quality plastic and a matching driver's license in his pocket, all bearing the name Stephen Leahy. What he didn't have was a knack for carding. He selected not one, not two, but four Anonimo watches, each worth between $1,000 and $3,000, and asked the store owner to ring each of them up separately on four different Visa cards, which he conspicuously pulled from a deck of a dozen. Two of the hefty transactions were declined, so Janer left with two watches worth a total of $5,777, charged to two Bank of America cards.

A patrol car pulled over Janer about two miles away. While the cops looked over Janer's genuine driver's license and asked him if he'd been watch shopping recently, a second cruiser drove by with the store owner in the passenger seat. He eyed Janer and confirmed that they had the right guy.

The cops arrested Janer and searched his car, pulling out the watches, twenty-eight credit cards, and six California driver's licenses, each with a different name. When detectives served a search warrant on his house they found more watches and a .22-caliber Walther P22 handgun.

The gun was bad news. Instead of a larceny charge and a probation violation Janer was now facing a federal beef for being a felon in possession of a firearm. Janer wasted no time in offering to lead the feds to the source of the counterfeit cards. In the standard arrangement for snitches, the government agreed to let Janer "proffer" his information under a limited grant of immunity: Nothing he said would be used directly against him. If they found it useful—if it led to arrests—they'd consider recommending a reduced sentence on his gun-possession charge.

In two proffer sessions totaling nearly eight hours, Janer spilled his guts to a local Secret Service agent and a federal prosecutor. He told them about Chris Aragon, his ring of cashers, and "Max the Hacker," a six-foot-five computer genius who'd been cracking banks from San Francisco hotel rooms.

He didn't know Max's last name, he said, but he'd once written a check to the hacker's girlfriend for $5,000. Her name was Charity Majors.

The Secret Service wrote up the interviews and entered the data into the agency's computer, but the agency never followed up on the information, and prosecutors declined to grant Janer any special consideration. He was sentenced to twenty-seven months in prison.

Max Vision had dodged a bullet. Janer's statements sank into a giant government computer—they might as well have been stashed in the cavernous warehouse in the final scene of *Raiders of the Lost Ark*. As long as nobody had occasion to dig them up, Max was safe.

Meanwhile, Max began the process of getting Carders Market up and running. He had plenty of experience setting up legitimate websites, but starting a crime site would take special preparations. For one thing, he couldn't just put the Carders Market server on the floor of his safe house—that would make him a sitting duck.

He hacked into a Florida data center run by Affinity Internet and installed a VMware virtual machine on one of its servers—secreting an entire simulated computer on one of its systems. His hidden server grabbed an unused Internet address from Affinity's pool of addresses. The site would be a ghost ship, not officially owned or operated by anyone.

Max played with different Internet forum software and finally settled on the flexible package vBulletin. He spent months customizing the layout and designing his own templates for the look and feel of the site, styling it in shades of gray and muted gold. The work felt satisfying. For the first time in years, he was creating something instead of stealing. It was just like setting up Whitehats.com, except in those ways in which it was the opposite.

Finally, on the one-year anniversary of the Operation Firewall raids, he conjured a new name in his ever-changing lineup of noms de guerre: Iceman. He chose the handle in part for its commonality: There were lots of Icemen in the underground—there'd even been one on Shadowcrew. If law enforcement tried to track him down, they'd find several mirages on their radar.

Max, as Iceman, launched Cardersmarket.com in late 2005 with little fanfare. Chris joined as the first coadministrator, inventing the handle EasyLivin' for the site.

From their careful observation of Shadowcrew and the splinter forums that followed, Max and Chris knew that the key to gaining acceptance was to appoint big names who could help run the board and attract still more heavy hitters from their circle of friends. The partners soon managed to draw two household names from the Shadowcrew diaspora.

Bradley Anderson, a forty-one-year-old Cincinnati bachelor, was their first pick. Anderson was a legend as "ncXVI," a fake-ID expert and author of the self-published book *Shedding Skin,* the bible of identity reinvention.

Their second recruit was Brett Shannon Johnson, thirty-five, a Charleston, South Carolina, identity thief famous online as "Gollumfun," a founder of both Counterfeit Library and Shadowcrew who'd retired from the latter site before the Secret Service swept in.

After vanishing from the scene for over a year, Johnson was crawling out of retirement—Chris's sidekick John Giannone had spotted him online that spring and struck up a conversation on ICQ, bringing him up to date on the latest busts and gossip.

Giannone wound up selling Johnson twenty-nine of Max's dumps for an easy six hundred bucks, then introduced him to Max, who sold him another five hundred cards. "I can see that you and I are going to be doing some good business in the future," Johnson had told Max.

Johnson accepted Max's and Chris's invitation to become an admin on Carders Market, lending the site the experience and contacts of the only Shadowcrew administrator to survive Operation Firewall.

Giannone joined Carders Market as "Zebra," and Max created a second, secret identity for himself, "Digits." The alternate handle was a keystone in Max's new business strategy. Shadowcrew had fallen because prosecutors proved that the founders were themselves buying, selling, and using stolen data—running an informational website wasn't, in and of itself, illegal, Max reasoned. So Iceman would be the public face of Carders Market but would never buy or sell stolen data. Digits, his alter ego, would handle that, vending the dumps Max was siphoning from the Vancouver pizza joint to anyone who could afford them.

To complete his vision for the site, Max needed one more admin with a particular qualification: a command of the Russian language. He wanted to repair the rift that Operation Firewall had torn between Eastern European carders and their Western counterparts. Two Russian Shadowcrew members had fallen into Cumbajohnny's VPN trap, and the whole affair had left the Russians deeply suspicious of English-speaking forums.

Max resolved that Carders Market would distinguish itself by having an Eastern European section moderated by a native Russian speaker. He just needed to find one.

Chris offered to help out, and Max accepted. If there was one thing that Chris had proven to his partner, it was that he knew how to recruit new talent.

20

The Starlight Room

Nine chandeliers hung over the lush velvet booths at Harry Denton's Starlight Room, the light scattering off a two-hundred-pound mirror ball suspended over the dance floor. Heavy crimson drapes parted from the picture windows like a stage, revealing the glimmering San Francisco skyline beyond.

Positioned on the twenty-first floor of the Sir Francis Drake Hotel, the Starlight Room was an opulent fixture in the city's teeming nightlife—a flashback to 1930s style, strewn with deep red and gold damask and hand-rubbed silk. More garish than hip, the club kept people coming by hosting regular theme nights. This was Russian Wednesday, and tuxedoed servers were pouring vodka shots at the crowded bar while music from the motherland spilled over the crowd.

In the ladies' room, Tsengeltsetseg Tsetsendelger was being kissed. Tipsy from a night out, the young Mongolian immigrant wasn't sure how it happened, or why, but a pretty five-foot-four girl with tumbling brown hair had decided to kiss her. Then Tsengeltsetseg blinked. There was another, identical woman beside her.

Michelle and Liz introduced themselves, and a wide, unaffected pumpkin smile crept onto Tsengeltsetseg's face. She told the Esquere twins that they could call her "Tea."

Tea was a regular at Russian Night and fluent in both Russian and

English. Born in northern Mongolia at a time when the country was still under Soviet influence, she'd learned Russian in school—until the Soviet empire collapsed and Mongolia's prime minister declared English the landlocked nation's official second language.

Looking for adventure and the proverbial better way of life, she won a student visa and emigrated to the United States in 2001. Her first thought upon landing at Los Angeles International Airport that summer was that Americans were awfully fat, but when she got out into the city she was more impressed; she enjoyed beautiful people, and L.A. was filled with them.

After one semester at a community college in Torrance, she moved to the Bay Area and got her green card. Now she was attending classes at Peralta College in Oakland, paying her rent and tuition by dishing ice cream at Fenton's Creamery.

Liz seemed strangely delighted to learn that Tea spoke Russian. The twins bought her a drink and then suggested they continue the party with some friends at their hotel four blocks away. It was after midnight when they got to Chris Aragon's suite at the luxe Clift Hotel near Union Square. Chris was relaxing there; Tea was struck at once by how handsome he was. He seemed interested in her as well, particularly after the twins mentioned that Tea knew Russian. Joined by two of Chris's female employees, they opened some booze and hung out until the small hours of the morning, when the girls all left to go to their own rooms and Tea crashed in Chris's for the night.

She was still shaking off sleep the next morning when the room became a hive of activity. Liz and a handful of other attractive young women—all alert and cleanly scrubbed after their night of partying—began popping in and out, receiving envelopes and cryptic instructions from Chris.*

* Liz was one of Chris Aragon's cashers, but there's no evidence that her sister Michelle was involved.

They came and went all day, picking up more envelopes, dropping off department store shopping bags, sometimes lingering for a time before departing again. The party atmosphere hung in the air, but there was a nervous, excited edge to it now that made Tea curious—but not so curious as to pry.

When the sun had set and the gang had gathered back at the suite, Tea said her good-byes; she had to go home to the East Bay, to be at work at the ice-cream parlor in the morning.

Chris had a better idea. He was starting a website with a business partner—"Sam"—and they happened to be in need of a full-time Russian translator. It would pay better than spooning out Coffee Cookie Dream to yuppies all day.

"Don't go," said Liz. "You'll make more money with us."

Tea looked over her pretty new friends. They reminded her of the New Russians who had emerged following the collapse of the Soviet regime, flush with suspiciously acquired wealth, consuming with more hunger than taste.

She liked Chris, though—he seemed different. And an Internet translating job would grant her the freedom and flexibility to focus on her college studies. She said yes.

The next day, Chris packed up his team for the next leg in their travels, a road trip to Vegas. Tea, he said, should meet them there for more fun. He told her to get a Yahoo! e-mail account, and he'd send her flight information once they'd arrived.

Back in her apartment, the whole adventure felt like a strange dream. But the next day, Tea had a confirmation number for her prepaid flight to Las Vegas in her Yahoo! in-box. She packed a bag and headed to the airport.

Chris relocated Tea to his own neighborhood and paid for her to rent an apartment in her real name in Dana Point, a coastal town in southern Orange County. At the end of a quiet, winding cul-de-sac, painted an Umbrian orange with Spanish tiles combing the roof, the "Tea House," as he dubbed it, was a world away from the Mongolian city where Tea grew up.

They made love on her new bed, and afterward, Chris left $40 on the nightstand so she could get her nails done. Tea's feelings were hurt. She wasn't a hooker. She was falling in love.

Chris and his team moved his card-printing gear from Villa Siena into the Dana Point apartment's attached garage—the Tea House would be his new plant and party house, as well as the base of operations for Tea's twenty-four-hour-a-day job on Carders Market. Her task would be to haunt the Eastern European carder forums, like Mazafaka and Cardingworld, and summarize what was happening there for the Russian section of Carders Market.

She'd need a "nick," Chris explained, a handle or nickname for her online alter-ego. She decided on "Alenka," the name of a Russian candy.

Alenka went to work at once, glued to the monitor at the Tea House day and night, doing her best to lure the high-powered Russians onto the site run by Chris and "Sam," the Whiz.

21

Master Splyntr

Taking up one floor of a lime-green office building on the bank of the Monongahela River, the National Cyber Forensics and Training Alliance was far removed from the cloistered secrecy of Washington's intelligence community, where Mularski had cut his teeth. Here, dozens of security experts from banks and technology companies worked alongside students from nearby Carnegie Mellon University in a cluster of neat cubicles, surrounded by a ring of offices that followed the smoked-glass walls around the building. With Aeron chairs and dry-erase boards, the office had the feel of one of the technology companies that provided the NCFTA with the bulk of its funds. The FBI had made a few changes before moving in, transforming one office into an electronic communications room, packed with government-approved computer and crypto gear to securely communicate with Washington.

In his office, Mularski looked over a "linkchart" Crabb, the postal inspector, had e-mailed him—a massive organization schematic showing the disparate connections among 125 hard targets in the underground. Mularski realized he'd been going about it all wrong by waiting for a crime, then working to track it back to the culprit. The criminals weren't hiding at all. They were advertising their services on the forums. That made them vulnerable, in the same way the New York and Chicago Mafia's rituals and strict hierarchy had given the FBI a roadmap to crack down on the mob decades before.

All he had to do now was join the carders.

He selected a forum from a list provided by Crabb and clicked on the account registration link. Under Justice Department regulations, Mularski could infiltrate the forums without approval from Washington, provided he observed strict limits on his activities. To maintain his cover, he could post messages to the forum bulletin boards, but he couldn't engage anyone directly; he would be permitted no more than three "substantive contacts" with any other forum member. Participating in crimes, or making controlled buys from a vendor, was out of the question. It could be an intelligence-gathering operation only; he would be a sponge, soaking up information about his adversaries.

As soon as he connected, he was confronted with his first important strategic decision: What would his hacker handle be? Mularski went with his gut. Inspired by the Saturday morning cartoon *Teenage Mutant Ninja Turtles,* the agent settled on the moniker of the sewer-dwelling karate champs' rodent sensei, a biped rat called Master Splinter. For uniqueness, and a hackerish timbre, he spelled his surname without major vowels.

So in July 2005, Master Splyntr signed up for his first crime forum, CarderPortal, laughing to himself over the poetry in assuming the name of an underground rat.

Mularski was soon playing the carder forums like a chessboard, drawing on the NCFTA's stream of scam data for his opening moves.

The center was plugged directly into the antifraud efforts at banks and e-commerce sites, so when a new criminal innovation showed up, Mularski knew about it. He posted about the schemes on CarderPortal, portraying them as his own inventions. The experienced crooks marveled at the newcomer who'd independently reinvented their newest tricks. And when the scams eventually became public in the press, the newbies remembered they'd heard it first from Master Splyntr.

In the meantime, the FBI agent was soaking up the history of the forums while honing his prose to affect the cynical, profanity-laced style of the underground.

After a few months, Mularski faced the first challenge to his intelligence-gathering operation. The initial crop of forums that grew from the detritus of Shadowcrew had been wide open to new members—spooked by Operation Firewall, many scammers had adopted new handles, and without reputations to trade on there'd been no way for carders to vet one another. Now that was changing. A new breed of "vouched" forums was emerging. The only way to get on them was to win the sponsorship of two existing members. Constrained by the Justice Department's guidelines, Mularski had deliberately avoided forming direct relationships in the underground. Who would vouch for him?

Borrowing a page from a Robert Ludlum novel, Mularski decided Master Splyntr needed a background legend that could propel him into the new crime boards. His thoughts turned to a Europe-based antispam organization called Spamhaus that he'd worked with as part of previous FBI initiatives.

Founded in 1998 by a former musician, Spamhaus charts the ever-changing lineup of Internet addresses spewing garbage into consumers' in-boxes; its database of spam sources is used by two-thirds of the world's ISPs as a blacklist. Of more interest to Mularski was the organization's public most-wanted list of notorious spammers. Peopled by the likes of Alan "Spam King" Ralsky and the Russian Leo "BadCow" Kuvayev, the Registry of Known Spam Operations, or ROKSO, is second only to a federal grand jury indictment on the list of places an Internet scammer doesn't want to see his name.

Mularski phoned up founder Steve Linford in Monaco to explain his scheme: He *wanted* to be on ROKSO—or, at least, he wanted Master Splyntr there. Linford agreed, and Mularski went to work crafting his background story. The best lies hew to the truth, so Mularski decided to make Splyntr a Polish spammer. Mularski was descended on his father's side from Polish immigrants—his bureau-issue button-down concealed a tattoo on his left arm of the Orzeł Biały, the white eagle with golden beak and talons that adorns Poland's coat of arms. Mularski would locate Master Splyntr in Warsaw; he'd visited Poland's capital and could roughly describe its landmarks if pressed.

In August, the ROKSO listing went live, for the first time stapling a "real" name to Mularski's cartoon-inspired alter ego.

> Pavel Kaminski aka "Master Splyntr" runs a loosely organized spam and scam crew from Eastern Europe. Possibly a BadCow affiliate. He is linked to: proxy spam; phishing; pump'n'dump; javascript exploits; carder forums; botnets.

The profile included samples of scammy spam messages supposedly sent out by "Pavel Kaminski," handcrafted by Spamhaus, and an analysis of his hosting arrangements.

Now the carders who Googled Master Splyntr could see for themselves that he was the real deal, a bona fide Eastern European cybercrook with sticky fingers in a lot of pies. When Mularski logged on to CarderPortal, he found business proposals waiting in his in-box from crooks hoping to partner with him. Still not allowed to engage any suspects, he blew them off sneeringly.

You're not much of a player, he'd write back. I don't want to deal with you because I'm a professional and you're obviously a newbie at this. To rebuff upper-echelon scammers, he challenged their pocketbooks: You don't have enough money to invest in what I'm doing.

Like an unattainable girl on prom night, Master Splytnr's aloofness only made him more attractive. When a new site called the International Association for the Advancement of Criminal Activity launched as a closed forum, he posted a simple note—Hey, I need a vouch—and two existing members spoke up for him solely on the strength of his reputation.

He was vouched on Theft Services next, then CardersArmy. In November 2005, he was one of the first members invited to a brand-new forum called Darkmarket.ws.

A few months later, another, competing site got big enough to cross his radar, and Master Splyntr joined Cardersmarket.com.

22

Enemies

Jonathan Giannone was learning that loss of privacy was the cost of doing business with Iceman.

He'd been working with the mystery hacker for over a year—mostly acquiring servers that Iceman used in his vulnerability scanning—and he was still constantly under Iceman's electronic scrutiny. One day, the hacker sent Giannone a link purporting to be a CNN article about computer problems at JetBlue, the airline that had rebuffed Giannone's long-ago extortion attempt. Giannone clicked on the link without thinking, and, just like that, Iceman was on his computer again. Client-side attacks for the win.

Giannone began routinely checking his computer for malware but couldn't keep up with Iceman's intrusions. Max got ahold of Giannone's United Airlines Mileage Plus password and began tracking his movements around the world—Giannone was a serious air travel aficionado who'd sometimes fly just to accumulate miles. When he'd land at San Francisco International, he'd find a text message from Iceman waiting for him on his cell. "Why are you in San Francisco?"

It might have been amusing if it weren't for Iceman's frightening mood swings. He could turn on you in a minute—one day you'd be his best friend, his "number one guy"; the next he'd be convinced you were a snitch, a ripper, or worse. He wrote Giannone long, unprompted e-mail diatribes, laundry lists of grievances against Chris or various members of the carding community.

It was jealousy, Giannone figured. While he and Chris were partying in Vegas and the OC, Iceman was locked in his apartment, working like a dog. Indeed, the hacker's outbursts often coincided with one of Giannone's California sojourns. In June 2005, Iceman picked a fight as Giannone boarded an early morning flight to Orange County—Iceman was taking him to task for some oversight in one of their joint operations. The first message hit Giannone's BlackBerry at six a.m.—three in the morning San Francisco time—and the texts continued nonstop for 2,500 miles before Iceman finally fell silent as the plane landed. When Giannone checked his e-mail later, he found dozens of apologetic letters from the hacker. "Sorry, I apologize. I was bugging out."

On an earlier occasion, in September 2004, Giannone told Iceman he was about to fly out to visit Chris, and Max remarked cryptically that he could prevent the trip if he wanted to. Giannone laughed. But an hour and a half into his flight, the plane suddenly turned around and headed for Chicago. As the airliner set down at O'Hare, the captain explained that the Los Angeles air traffic control center had gone dark, necessitating the change in itinerary.

It turned out a computer error was responsible. There was a known bug in the Windows-based radio control system at the Los Angeles Air Route Traffic Control Center in Palmdale, which required technicians to reboot the machine every 49.7 days. They'd missed a reboot, and a backup system had failed at the same time. The outage resulted in hundreds of flights being grounded and five incidents of airplanes drifting closer to each other than safety regulations permit. No foul play was discovered, but years later, when the full range of Max Vision's powers became clear, Giannone would find himself wondering if Iceman hadn't cracked the FAA's computers and crippled Los Angeles, just to stop him from going clubbing with Chris.

Giannone finally took radical measures to try to keep Iceman out of his stuff: He bought an Apple. Iceman could penetrate just about anything. But Giannone was pretty sure he couldn't hack Macs.

While Max kept up surveillance of his crime partners, Carders Market began slowly generating buzz, intensified by the mysterious swagger of its founders. As Iceman and Easylivin', Max and Chris were unknown quantities among their fellow crooks, but experienced carders could practically smell the confidence and street smarts in their posts.

In Seattle, word of the new site reached Dave "El Mariachi" Thomas, the former FBI asset who, like Max, had tried to blow the whistle on Operation Firewall. Thomas had been feeling adrift since the feds pulled the plug on his intelligence-gathering operation, and he was looking for a new online home.

Wary at first, Thomas registered under a fake handle. But when Iceman invited a public discussion of Carders Market's philosophy and charter, Thomas dove in, opining in detail on the course the site should follow to nurture successful ops while avoiding Shadowcrew's fate.

At first, Chris and Max thought Thomas might be a valuable contributor. But they soon detected that he had a beef with one of their handpicked admins, Brett "Gollumfun" Johnson.

Rumors had been swirling about Johnson since his return to the scene—you don't just disappear for two years and then come back onto the carder forums as though nothing has happened. In August, a hacker called "Manus Dei"—the Hand of God—added fuel to the fire when he cracked Johnson's e-mail account and posted a blistering profile of the carder on a Google Group called FEDwatch. The write-up gave Johnson's real name, his current address in Ohio, and a slew of personal details stolen from his in-box. Among the revelations: Johnson had been corresponding with a *New York Times* reporter about the carding scene and had registered a mysterious domain name, Anglerphish.com—perhaps in preparation for starting his own site.

There was nothing to suggest that Johnson was snitching, though, and neither Max nor Chris had been particularly alarmed by the info dump. Thomas, on the other hand, was now convinced the Shadowcrew founder was an informant. After all, Johnson had announced his retirement before Operation Firewall and then reappeared afterward with no real explanation.

The last thing Chris and Max needed on their emerging site was a shootout between two old-school carders with a Shadowcrew-era grudge. Still possessed by an entrepreneurial pride, Chris wanted the site to be the best crime forum possible. So he reached out to Thomas by ICQ to try to head off trouble.

"I'm not going to entertain any drama about Gollumfun, or others, who is a rat who isn't a rat," Chris wrote. "I just want a clean nice board so we can have a safe place to play."

Chris promised he'd give Johnson the same message: Play nice. It was Conflict Resolution 101. He followed the paternalistic lecture by asking Thomas's advice on running a successful forum—showing the elder carder respect for his years of experience. But to make sure his admonition was taken seriously, Chris added a warning. "We are not kids dude," he wrote. "We are very old school. And we are very good at what we do."

Thomas promised to behave, adding that he'd do his best to help make Carders Market the drama-free forum everyone wanted. But secretly, a hard pit of suspicion was forming in his gut. Why would anybody defend Brett Johnson, who was so obviously a snitch?

He noticed that Easylivin' was using an old version of ICQ that leaked an Internet IP address. Thomas tried to trace the address and wound up in Boston, a known hotbed of federal informants. Carders Market's hosting was based in Ft. Lauderdale, Florida, another perfect place to run an undercover operation. And the phone number on the domain name listing went to a police department in California, albeit in a different area code. That was probably a coincidence, but who knows?

When he was done adding up the evidence, he felt sick to his stomach. Carders Market was a federal sting. It was obvious now. He vowed to himself that he'd do everything he could to destroy the new site and bring down the old-school assholes Easylivin' and Iceman.

23

Anglerphish

Max was developing suspicions of his own about Brett Johnson. He began keeping a close eye on the admin on Carders Market, checking his access logs and scouring his private messages. For good measure, he hacked into Johnson's account on the International Association for the Advancement of Criminal Activity, IAACA, and reviewed his activity there. He found no smoking gun.

Could he really have brought an informant into the inner circle of his new crime site?

The problem was that there was no reliable test to determine if Johnson, or anyone else, was working for the government. Max wanted one badly—a jurisprudence security hole, like the buffer overflow in BIND, that he could use over and over again on anyone he suspected. `If (is_snitch(Gollumfun)) ban(Gollumfun);`. He confided in David Thomas, not realizing that Thomas had already put Iceman on his mile-long enemies list.

```
At one point in checking him out, he sent us some
PayPal fulls that were valid, which I pegged as
illegal. It made me think, okay, this guy isn't a
fed or fed lackey.
```

This is very important for me to find out, because it is how I have been making trust decisions. We have it in mind to have a lawyer give us the definitive answer, my partner said he was on that and would find out. I am skeptical that we'll ever get a straight answer though, because lawyers seem to enjoy taking your money and providing you heuristic guesses rather than concrete facts. Maybe I've just had bad lawyers.

I would really like to know a specific way that I can find something a cop or CI can't do. Something that if they do it, their cases are all thrown out 100%. What a holy grail. So far I have been living as though "doing a criminal act" disqualifies them. Like people who smoke a joint with someone to make sure that person isn't a cop. Or a hooker who asks her john, "Are you a cop? You know you have to tell me if you are."

Brett Johnson was indeed dirty. But contrary to suspicions, his return to crime in the post-Firewall era hadn't started as a snitching expedition. It had all begun with a girl.

Johnson's crime and cocaine habits had driven away his wife of nine years—she threw out his MSR206 on her way out the door—and he'd been seeing a psychologist to cope with the loss. Then he met Elizabeth in a North Carolina bar. She was a twenty-four-year-old exotic dancer at a local strip club, and for Johnson it was love at first sight. He burned through his savings to buy her gifts, a $1,500 purse here, a $600 pair of shoes there, and she moved in with him after five months. But when they had sex for the first time, she wouldn't let him kiss her.

Johnson's darkest suspicions were confirmed when he located Elizabeth on a website on which men post reviews of strippers and prostitutes. There it was, line after line of disgusting detail about the services his girlfriend had been providing in exchange for cocaine and cash. He confronted her with the evidence, and she tearfully promised to quit the drugs and the prostitution.

Hoping to wrench her from the patterns of her old life, Johnson showered Elizabeth with more gifts and expensive dinners out. It was that, and not any hidden agenda, that impelled his return from retirement. He needed the money, plain and simple.

The luck that had seen him through Operation Firewall failed him on February 8, 2005, when Charleston, North Carolina, police busted him for using counterfeit Bank of America cashier's checks to pay for Krugerrands and watches he won on eBay and had shipped COD to his drops. After a week of stewing in the Charleston County Detention Center, pining for Elizabeth, the Secret Service paid him a visit. Once he convinced them he was Gollumfun—the admin who got away when they dropped the hammer on Shadowcrew—they agreed to help him with his state case if he'd work for them.

The Secret Service had Johnson's bail lowered to $10,000. When he bonded out, the agents moved him from Charleston to Columbia, South Carolina, where they rented him a corporate apartment and paid him a $50 per diem. Now he was a daily visitor to the Columbia field office, checking in at four p.m. and working until nine, taking the Secret Service deep into Carders Market and the other boards. Everything that crossed his computer was recorded and displayed simultaneously on a forty-two-inch plasma screen hanging on the wall of the office.

They called it Operation Anglerphish, and Johnson thought it would make a great book one day. That's why he'd registered the domain name Anglerphish.com and opened up talks with a *New York Times* reporter. When Manus Dei cracked his e-mail and revealed those activities online, Johnson's Secret Service handlers were irate. They promptly banned him from using computers away from the office and told him to cut off contact with the reporter. Elizabeth left him—her name and occupation had been exposed in the breach.

Then Iceman stripped Johnson of his privileged position in Carders Market, and crooks he'd known since the Counterfeit Library days started refusing to do business with him. Johnson was running out of credibility, and the Secret Service was running out of patience.

In late March 2006, the agents decided to act on one of Anglerphish's only catches, a California identity thief who'd stolen at least $200,000 by e-filing bogus tax returns through H&R Block, then collecting the refunds himself. Johnson, an expert in that particular scam, had been talking with the crook online, and the Secret Service had traced the chats to the C&C Internet Café in Hollywood. A Los Angeles agent visited the coffee shop and sat two tables away while the man filed his fake returns.

But when local police and Secret Service agents raided the target's Hollywood apartment, they found it had been cleaned out: no computers and not a shred of documentary evidence. The suspect had done everything but deep-clean the carpet and paint the walls.

Johnson's handlers in Columbia already suspected their asset of leaking his informant status after the drama on Carders Market. Now they had reason to believe he'd tipped off the target of an impending raid. They brought in a polygraph examiner and strapped Johnson to the box.

The needles were steady as Johnson answered the first two questions: Did he contact the target? Did he have anyone else contact the target? No and no. The final question was broader: Did Johnson have any unauthorized contact with anyone? "No," he said again, his galvanic skin response skittering up the chart.

Despite the agents' admonishments, Johnson had secretly continued his talks with the *New York Times* reporter, he admitted, and he was very serious about getting a book deal. The feds interrogated him until two in the morning, then had him sign a form consenting to a search of his agency-funded apartment.

Tossing the apartment was like an Easter egg hunt. The agents found a stored value card in a shoe in the bedroom closet. A memo book containing account numbers, PINs, and identity information was in a toiletry kit in the bathroom. A sock stuffed in a pair of men's pants in the closet contained sixty-three ATM cards. A Rubbermaid bowl at the bottom of the laundry bin was keeping fresh nearly two thousand dollars in cash. Finally, there were loaded Kinko's payment cards; Johnson had been buying computer time at the local copy shop.

He'd been leading a triple life almost from the start of his service to the agency, posing as a crook at the Columbia field office and pulling his own very real capers in his off hours.

Johnson's specialty was the same scam the Los Angeles target had been carrying out. He'd mine victims' Social Security numbers from online databases, including California's Death Index of recently departed Golden State residents, then file bogus tax returns on their behalf, directing the refunds into prepaid debit cards that could be used for ATM withdrawals. He'd pulled in more than $130,000 in tax refunds under forty-one names, all under the nose of the Secret Service.

The agents phoned up Johnson's bail bondsman and persuaded him to revoke the $10,000 bond that had set the fraudster free. Then they put Johnson back in the county jail. After three days, Johnson's handler showed up with a senior agent, who was not happy with the informant. "Before we begin, Brett, I just want to say that you are either going to tell us everything that you have done the past six years, or I'm going to make it my mission in life to fuck over you and your family," the supervisor growled. "And I'm not just talking about these current charges. Once you get out, I will hound you for the rest of your life."

Johnson refused to cooperate, and the agents stormed out. The U.S. Attorney's Office started working on a federal indictment. But the swindler had one more trick up his sleeve. Two weeks later he managed to get his bond reinstated, bailed from the detention center, and promptly vanished.

Anglerphish was a debacle. After 1,500 hours of work, the government was left with a fugitive informant and tens of thousands of dollars in new fraud. There was only one silver lining: that first batch of twenty-nine platinum dumps Johnson had bought in May for $600.

The Secret Service had tracked some of the cards to a pizza parlor in Vancouver—a dead end. But the corporate Bank of America account the seller used to accept his payment belonged to one John Giannone, a twenty-one-year-old living in Rockville Centre on Long Island.

24

Exposure

"Tea, these girls are white trash. Don't be friends with them," said Chris. "Their minds are different."

They were at Naan and Curry, a twenty-four-hour Indian and Pakistani restaurant in San Francisco's theater district. It had been three months since Tea hooked up with Chris, and she was with him for one of his monthly trips to the Bay Area, where'd he'd meet his mysterious hacker friend "Sam" just before dawn. They were only four blocks from Max's safe house now, but Tea wouldn't be introduced to the hacker on this trip or any other. Nobody met Sam in person.

She was fascinated by how it all worked: the cashless nature of the crime, the way Chris organized his crew. He'd told her everything, once he thought she was ready, but never asked her to hit the stores with the others. She was special. He didn't even like her hanging out with his cashing crew, for fear that they'd somehow taint her personality.

Tea was also the only employee not being paid. After she'd protested the $40 Chris left on the nightstand, Chris concluded that Tea didn't want any money from him at all, despite the long hours she was spending on Carders Market and the Russian crime boards. Chris was taking care of the rent on the Tea House, buying her clothes, and paying for her travel—but she found it a strange existence, living online, traveling on confirmation numbers instead of plane tickets. She'd become a ghost, her body in

Orange County, her mind more often projecting into Ukraine and Russia, befriending organized cybercrime chieftains in her role as Iceman's emissary from the carding world of the West.

Iceman, she'd decided, was pretty cool. He was always respectful and friendly. When Chris and his partner got into one of their fights, each man would whine and gossip about the other to Tea over ICQ, like children. At one point, Iceman sent her a bunch of dumps and suggested she go into business for herself, a move that sent Chris into a petulant rage.

As Chris and Tea chatted over Indian food, a tall man with a ponytail walked in from the street and headed for the cash register in back, his eyes flickering over them, just for a moment, before he picked up a bag of takeout and left.

Chris smiled. "That was Sam."

Back in Orange County, Chris's counterfeiting operation was earning enough for him to send his kids to private schools, cover Tea's apartment, and, in July, start searching for a bigger and better home for himself and his family. He went house-hunting with Giannone and found a spacious rental—a two-story house in the coastal town of Capistrano Beach at the end of a quiet cul-de-sac on a bluff rising above the sandy beach. It was a family-friendly neighborhood, basketball hoops hanging above garages and a boat parked in a neighbor's driveway. His move-in date was July 15.

Giannone flew back out for the July 4 weekend—Chris's last holiday at his old condo—but wound up back at the Tea House while Chris spent time with his family. It happened all the time; Giannone would fly into John Wayne Airport, expecting a weekend of clubbing with Chris, and instead would end up holed up with one of the crew or be tasked with babysitting Chris's boys at his house. Tea was tolerable, different from the cheap party girls cashing out Chris's cards, but time at the Dana Point apartment dragged.

He phoned Chris and complained that he was bored. "Come to the house," Chris said. They were at the pool. "The wife's here with the kids."

Giannone invited Tea, who'd never seen Chris's condo complex just four miles away. When they arrived, Chris, Clara, and the two boys were splashing around in the pool, enjoying the sun. Giannone and Tea said hello and made themselves at home on some deck chairs.

Chris looked stunned. "I see you brought your friend," he said to Giannone testily.

Clara knew Giannone, the babysitter, but had never met Tea. She looked at the stranger, then at Giannone, then back at the Mongolian, awareness and anger creeping over her face.

Giannone realized he'd made a blunder. The two women looked uncannily alike. Tea was a younger version of Chris's wife, and at a glance, Clara knew her husband was sleeping with this woman.

Chris pulled himself out of the pool and walked around to where they were sitting, his face neutral. He squatted down in front of Giannone, his hair dripping water onto the concrete. "What are you doing?" he said in a low voice. "Get out of here."

They left. And for the first time since she joined up with Chris Aragon and his gang, Tea felt dirty.

Chris wasn't angry—he got a guilty, alpha-male pleasure out of seeing Tea and Clara in the same place. But Tea's crush was becoming a problem. He had genuine affection for her and her quirky ways, but she was becoming an unwanted complication.

There was an ideal solution at his disposal. He bought her a plane ticket to visit her home country for an extended vacation, literally banishing his overardent paramour to Outer Mongolia.

With Chris distracted by his tangled love life, Carders Market was consuming more of Max's time, and he still had his business as "Digits" to run. He was working in the food service industry now, and it was paying off big.

It had started in June 2006, when a serious security hole emerged in the software RealVNC, for "virtual network console"—a remote-control program used to administer Windows machines over the Internet.

The bug was in the brief handshake sequence that opens every new session between a VNC client and the RealVNC server. A crucial part of the handshake comes when the server and client negotiate the type of security to apply to the session. It's a two-step process: First, the RealVNC server sends the client a shorthand list of the security protocols the server is configured to support. The list is just an array of numbers: [2,5], for example, means the server supports VNC's type 2 security, a relatively simple password authentication scheme, and type 5, a fully encrypted connection.

In the second step, the client tells the server which of the offered security protocols it wants to use by sending back its corresponding number, like ordering Chinese food off a menu.

The problem was, RealVNC didn't check the response from the client to see if it was on the menu in the first place. The client could send back any security type, even one the server hadn't offered, and the server unquestioningly accepted it. That included type 1, which is almost never offered, because type 1 is no security at all—it allows you to log in to RealVNC with no password.

It was a simple matter to modify a VNC client to always send back type 1, turning it into a skeleton key. An intruder like Max could point his hacked software at any box running the buggy RealVNC software and instantly enjoy unfettered access to the machine.

Max started scanning for vulnerable RealVNC installations as soon as he learned of this gaping hole. He watched, stunned, as the results scrolled down his screen, thousands of them: computers at homes and college dorms; machines in Western Union offices, banks, and hotel lobbies. He logged in to some at random; in one, he found himself looking at the feeds from closed-circuit video surveillance cameras in an office-building lobby. Another was a computer at a Midwest police department, where he could listen in on 911 calls. A third put him in a home owner's climate control system; he raised the temperature ten degrees and moved on.

A tiny fraction of the systems were more interesting and also familiar from his ongoing intrusion into the Pizza Schmizza: They were restaurant point-of-sale systems. They were money.

Unlike the simple dumb terminals sitting on the counters of liquor stores and neighborhood grocers, restaurant systems had become sophisticated all-in-one solutions that handled everything from order taking to seating arrangements, and they were all based on Microsoft Windows. To support the machines remotely, service vendors were installing them with commercial back doors, including VNC. With his VNC skeleton key, Max could open many of them at will.

So Max, who'd once scanned the entire U.S. military for vulnerable servers, now had his computers trolling the Internet day and night, finding and cracking pizza joints, Italian *ristorantes*, French bistros, and American-style grills; he harvested magstripe data everywhere he found it.

Under Visa-issued security standards, that shouldn't have been possible. In 2004 the company outlawed the use of any point-of-sale system that stores magstripe data after a transaction is complete. In an effort to comply with the standards, all the major vendors produced patches that would stop their systems from retaining the swipes. But restaurants weren't racing to install the upgrade, which in some cases was a paid extra.

Max's scanning machinery had several moving parts. The first was aimed at finding VNC installations by performing a high-speed "port sweep"—a standard reconnaissance technique that relies on the Internet's openness and standardization.

From the start, the network's protocols were designed to let computers juggle a variety of different types of connections simultaneously—today that can include e-mail, Web traffic, file transfers, and hundreds of other more esoteric services. To keep it all separate, a computer initiates new connections with two pieces of information: the IP address of the destination machine, and a virtual "port" on that machine—a number from 0 to

65,535—that identifies the type of service the connection is seeking. The IP address is like a phone number, and a port is akin to a telephone extension you read off to the switchboard operator so he can send your call to the right desk.

Port numbers are standardized and published online. E-mail software knows to connect to port 25 to send a message; Web browsers connect to port 80 to retrieve a website. If a connection on the specified port is refused, it's like an unanswered extension; the service you're looking for isn't available at that IP address.

Max was interested in port 5900—the standard port for a VNC server. He set his machines sweeping through broad swaths of Internet address space, sending to each a single sixty-four-byte synchronization packet that would test whether port 5900 was open for service.

The addresses that answered his sweep streamed into a PERL script Max wrote that connected to each machine and tried to log in through the RealVNC bug. If the exploit didn't work, the script would try some common passwords: "1234," "vnc," or an empty string.

If it got in, the program grabbed some preliminary information about the computer: the name of the machine and the resolution and color depth of the monitor. Max snubbed computers with low-quality displays, on the assumption that they were home PCs and not businesses. It was a high-speed operation: Max was running on five or six servers at once, each capable of zipping through a Class B network, over sixty-five thousand addresses, in a couple of seconds. His list of vulnerable VNC installations grew by about ten thousand every day.

The point-of-sale systems were needles in a massive haystack. He could spot some just from the name: "Aloha" meant the machine was likely an Aloha POS made by Atlanta-based Radiant Systems, his favorite target. "Maitre'D" was a competing product from Posera Software in Seattle. The rest of them took some guesswork. Any machine with a name like "Server," "Admin," or "Manager" needed a second look.

Slipping in over his VNC client, Max could see what was on the computer's screen as though standing right in front of it. Since he worked at night, the display on the dormant PC was usually dark, so he'd nudge his mouse to clear the screen saver. If there was anyone in the room, it might have been a little spooky: Remember that time your computer monitor flipped on for no reason, and the cursor twitched? It might have been Max Vision taking a quick look at your screen.

That manual examination was the slow part. Max recruited Tea to help out—he gave her a VNC client and started feeding her lists of vulnerable machines, along with instructions on what to look for. Soon, Max was wired into eateries throughout America. A Burger King in Texas. A sports bar in Montana. A trendy nightclub in Florida. A California grill. He moved up to Canada and found still more.

Max had gotten his start vending by stealing the dumps from a single restaurant. Now he had as many as a hundred feeding him credit card data in nearly real time. Digits would be doing a lot more business.

With so much work to be done, Dave "El Mariachi" Thomas had chosen a bad time to become a real pain in Iceman's ass. In June, Thomas did something nearly unheard of in the insular computer underground: He took their dispute off the forums and into public, civilian cyberspace, attacking Carders Market in the comments section of a widely read computer security blog, where he accused Iceman of being "LE"—law enforcement.

"Here is a site hosted in Ft Lauderdale Florida," Thomas wrote. "Matter of fact, it's hosted right out of a guy's house. Yet, LE refuses to shutter them. Instead, this site promotes vending of PINs and numbers and PayPals and eBays and so forth, all the while LE looks on at all the players.

"LE claims they can't do anything to a site hosted on U.S. soil. Yet, truth be told, it's LE running the site just like they ran Shadowcrew."

By highlighting Carders Market's hosting arrangements, Thomas was targeting Iceman's Achilles' heel. The site had been purring along unmolested because Affinity didn't notice the illicit server among its tens of thousands of legitimate hosted sites. El was working to change that, lodging complaints with the company over and over again. The tactic was lacking in logic: If Carders Market really was under government control, the complaints would fall on deaf ears; only if it was a real crime site would Affinity kick it off. If Iceman drowns, then he's not a witch.

A week after Thomas's post, Affinity abruptly cut off Carders Market. The shutdown angered Max; he'd had a good thing going at ValueWeb. He searched overseas for new, legitimate hosting that would stand up to El Mariachi, approaching companies in China, Russia, India, and Singapore. It always turned out the same way—they'd demand some upfront money as the price of admission and then roll a spool of red tape in front of the door, asking for a passport and a business license or corporate papers.

"Couldn't be because you have some STUPID FUCKING NAME called CARDERS this or CARDERS MARKET that, now could it?" Thomas wrote, taunting Iceman. "Maybe if you didn't scream 'CARDERS WORK HERE,' you could get a small site going, and possibly grow to be the beast you so desperately need to be."

It was personal now: Thomas hated Iceman, whether he was a fed or not, and the feeling had become mutual.

Max finally set up at Staminus, a California firm specializing in high-bandwidth hosting resistant to DDoS attacks. By then, Thomas was tearing into him in the comments section of a random blog called "Life on the Road." The blogger had quoted Thomas's comments about Carders Market in a brief entry about the forums, unwittingly volunteering his blog as the new battlefield in the El Mariachi-versus-Iceman war.

Iceman picked up the gauntlet and posted a lengthy public rebuttal to Thomas's indictment, accusing his foe of "hypocrisy and slander."

```
CM is NOT a "crime board" or an "empire" or
any of this bullshit accusation. We are simply
a forum that chooses to allow discussion of
financial crime. We also lend authority in judging
```

> which members are real and which are the fakes, but those are just our opinions, we make no money from this service. We are just a CARRIER for the information, a FORUM through which this communication can occur without oppression. CM is not involved in any crime whatsoever. It is not illegal to operate a forum and allow discussion.
>
> Craigslist.com has people posting about prostitution, drug hookups, and other obvious crime, yet people don't call craigslist a "hookers and blow one stop shop" or a crime empire. It is recognized as a CARRIER which is not responsible for the content of posts therein. This is the state of Carders Market.

The spirited defense completely ignored the detailed crime tutorials and review system on Carders Market, not to mention the secret impetus for the site: to give Max a place to sell stolen data.

Knowing his California hosting wouldn't satisfy the underground, Max resumed his search for an arrangement overseas. The next month, he hacked himself a new server, this time in a country as far from U.S. influence as any on the Net—a nation unlikely to respond to complaints from Dave Thomas or even the American government.

"Carders Market is now hosted in IRAN," he announced on August 11. "Registration is reopened."

25

Hostile Takeover

"Rapidity is the essence of war. Take advantage of the enemy's unreadiness, make your way by unexpected routes, and attack unguarded spots."

Max had been reading Sun Tzu's *The Art of War,* using the 2,600-year-old tome as his hacking manual. He sketched out his plans on a pair of whiteboards in his safe house; after some attrition and new entrants, there were five English-language carding sites that mattered in the underground, and that was four too many. He'd spent weeks infiltrating his competitors: ScandinavianCarding, the Vouched, TalkCash, and his chief rival, DarkMarket, the UK-run site that emerged a month before Carders Market and was building a powerful reputation as a ripper-free zone.

In a way, Max's plan to muscle in on the other forums was coming from the white-hat side of his personality. The status quo was working fine for Max the criminal—he wasn't greedy, and he was doing brisk business on Carders Market. But the post-Shadowcrew carding scene was broken, and when Max the white hat saw something broken, he couldn't resist fixing it—just as he'd done for the Pentagon a few years earlier.

Ego played a role too. The whole carding world seemed to think Iceman was just another forum administrator, bankrupt of any skill except the ability to set up forum software. Max saw a golden opportunity to show the carders how wrong they were.

DarkMarket turned out to be an unguarded spot. A British carder called JiLsi ran the site, and he'd made the mistake of choosing the same password—"MSR206"—everywhere, including Carders Market, where Max knew everyone's passwords. Max could just walk in and take over. The Vouched, on the other hand, was a fortress—you couldn't even connect to the website without a privately issued digital certificate installed in your browser. Fortunately, JiLsi was also a member of that site, and he had moderator privileges there. Max found a copy of the certificate in one of JiLsi's webmail accounts, protected by the carder's usual password. From there, it was just a matter of logging in as JiLsi and leveraging his access to get at the database.

On TalkCash and ScandinavianCarding, Max determined that the forum software's search function was vulnerable to an "SQL injection" attack. It wasn't a surprising discovery. SQL injection vulnerabilities are the Web's most persistent weakness.

SQL injection has to do with the behind-the-scenes architecture of most sophisticated websites. When you visit a website with dynamic content—news articles, blog posts, stock quotes, virtual shopping carts—the site's software is pulling the content in raw form from a back-end database, usually running on a completely different computer than the host to which you've connected. The website is a facade—the database server is the important part, and it's locked down. Ideally, it won't even be accessible from the Internet.

The website's software speaks to the database server in a standard syntax called Structured Query Language, or SQL (pronounced "sequel"). The SQL command SELECT, for example, asks the database server for all the information that fits a specified criteria. INSERT puts new information in the database. The rarely used DROP instruction will mass-delete data.

It's a potentially perilous arrangement, because there are any number of situations where the software has to send a visitor's input as part of an

SQL command—in a search query, for example. If a visitor to a music site enters "Sinatra" in the search box, the website's software will ask the database to look for matches.

```
SELECT titles FROM music_catalog
    WHERE artist = 'Sinatra';
```

An SQL injection vulnerability occurs when the software doesn't properly sanitize the user's input before including it in a database command. Punctuation is the real killer. If a user in the above scenario searches on "Sinatra'; DROP music_catalog;" it's tremendously important that the apostrophe and semicolons not make it through. Otherwise, the database server sees this.

```
SELECT * FROM music_catalog
WHERE artist = 'Sinatra'; DROP music_catalog;';
```

As far as the database is concerned, that's two commands in succession, separated by a semicolon. The first command finds Frank Sinatra albums, the second one "drops" the music catalog, destroying it.

SQL injection is a standard weapon in every hacker's arsenal—the holes, even today, plague websites of all stripes, including e-commerce and banking sites. And in 2005, the forum software used by TalkCash and ScandinavianCarding was a soft target.

To exploit the bug on TalkCash, Max registered for a new account and posted a seemingly innocuous message on one of the discussion threads. His SQL attack was hidden in the body of the message, the font color set to match the background so nobody would see it.

He ran a search query designed to find the post, and the buggy forum software passed his command to the database system, which executed it, INSERTing a new administrator account just for Max. A similar attack worked at ScandinavianCarding.

On August 14, Max was ready to show the carding world what he was capable of. He slid into the sites through the holes he'd secretly blasted in their ramparts, using his illicit admin access to copy their databases. The plan would have made Sun Tzu proud: Attacking and absorbing rival forums was an unexpected route indeed. Most carders wanted to avoid attention, not thrust themselves into prominence. A hostile takeover was unprecedented.

When he was done with the English-speaking sites, Max went to Eastern Europe. He'd strived to unite the Eastern European carders with the West, but Tea's efforts had been largely fruitless—the Russians liked her but didn't trust an American board. Diplomacy had failed; it was time for action. He found Cardingworld.cc and Mazafaka.cc no more secure than the western boards and was soon downloading their databases of private messages and forum posts. Megabytes of Cyrillic flowed onto his computer, a secret history of scams and hacks against the West stretching back months, now permanently warehoused on Max's hard drive in San Francisco's Tenderloin.

When he was done, he executed the DROP command on all the sites' databases, wiping them out. ScandinavianCarding, the Vouched, Talk-Cash, DarkMarket, Cardingworld—the bustling, twenty-four-hour-a-day marketplaces supporting a billion-dollar global underground economy all winked out of existence. Ten thousand criminals around the world, men with six-figure deals in the works; wives, children, and mistresses to support; cops to buy off; mortgages to pay; debts to satisfy; and orders to fill, were, in an instant, blind. Adrift. Losing money.

They would all know the name "Iceman."

Max then went to work on the stolen membership data, ignoring, for now, the Eastern European carders. After culling the duplicates and undesirables from the four English-language sites, there were 4,500 new members for Carders Market. He rolled them all into his site's database, so the carders could use their old nicknames and passwords to log in to their new home. Carders Market had six thousand members now. It was larger than Shadowcrew had ever been.

He announced the forced merger in a mass e-mail to his new members. As the morning dawned in San Francisco, he watched them gather, confused and angry, on his consolidated crime forum. Matrix001, a German DarkMarket administrator, demanded an explanation for Iceman's actions. A previously taciturn spam king named Master Splyntr spoke up to criticize the organization of the material Iceman had stolen from the other boards. The entire contents of the competing sites now lived in a new section of Carders Market called "Historical posts from merged forums." They were unsorted and difficult to navigate; Max had found the sites' content worthy of preserving but not of organizing.

Max watched the grumbling for a while, then stepped in and let everyone know who was in charge.

> @Master Splyntr: unless you have something constructive or specific to say, your comment is unwelcome. If you are unhappy with the layout, then go away and come back later, because it is not yet sorted out!
>
> @matrix001: The old forums were negligent in their security, using shared hosting, failing to use encryption of the data, logging IP addresses, using "1234" as the administrative passwords (yes really people this is true!), and general administrative Nazism. Some, such as TheVouched, were even giving a false sense of security, which as you know is far worse than none at all.
>
> You ask, what is the meaning of "all this"? If you mean, why would we merge five carding forums together, the short answer is because I didn't have time nor interest to merge in the other four for a total of nine!

> Basically, this was overdue. Why have five different forums each with the same content, splitting users and vendors, and a mish-mash of poor security and sometimes poor administration and poor moderation. I am not saying that is the case in all, but it was for most.
>
> With the right moderation, CM will return to its previous "tight" reign, with zero tolerance policy against ripping, and almost anarchist policy of not locking threads and promoting discussion. In the meantime, there is extra "fluff" from the previous forums, but that will be cleaned up.
>
> What is the point? Security. Convenience. Increase quality and decrease the noise. Bringing order to a mess . . .

A Canadian hacker called Silo countered that Iceman had dissolved the social glue that held the carder community together. He'd violated their trust.

> You breached our community's security. Stole the databases of other forums. Couldn't your merger have taken place with the admins of all the boards consenting to it? What's the difference between me hacking your e-mails and reading up on your business and posting your communications on my board?
>
> Either way you look at it, you've breached what little trust exists in the community. My suggestion is that you delete the databases you

have that aren't yours to display. The proper thing to do is ASK the admins of the boards if one true unified board is in the best interests of our community, and wait and see if they would be interested in such a board.

That is my two cents.

There are people out here with a lot of skills Iceman. How they use them is what determines our community.

The Vouched came back online, but not for long—it was supposed to be a private, secure forum open only to a select few. When Max had broken its security, he'd shattered its credibility, and nobody bothered to return. TalkCash and ScandinavianCarding were doomed—they had no backups of the databases Max had destroyed. Their members mostly stayed on at Carders Market.

Aside from the Russian forums, which Max was having trouble assimilating because of the language barrier, there was just one black mark on Max's triumph: DarkMarket. His chief competitor had backups and managed to crawl back to life within days. It was a slap in the face to everything Max was trying to achieve for himself and the community. The war had begun.

In Orange County, Chris was consolidating his end of the business too. He decided it would be convenient to have his full-time workers all living in the same place, and the Archstone chain of apartment complexes offered an Internet-based move-in process perfectly suited to his plans. Prospective tenants could fill out a lease on the company's website and pay the easy $99 deposit and the first month's rent with a credit card. Chris could handle everything online, and his people wouldn't have to put in an appearance until move-in day, when they'd stop by the rental office to flash their fake ID and pick up the door key.

He moved two of his cashers, and Marcos, his pot connection, into the Archstone Mission Viejo, a labyrinth of McMansion-style apartments painted the colors of a sunset and clinging to a hill dotted with palm trees and high-tension lines alongside Interstate 5, ten minutes from his house. He was also looking to expand his crew. One girl had dropped out and moved to Toledo after her second in-store bust, and two others had quit in disgust when Chris impregnated his teenage girlfriend—he was now paying for an apartment for the young woman and their son, whose existence he kept secret even from his mother.

At the NCFTA office in Pittsburgh, Keith Mularski, in his Master Splyntr guise, got a private message from Iceman himself two days after the hostile takeover. The hacker wanted to apologize for some of his hasty words on his forum.

Anticipating the next stage in the DarkMarket–Carders Market conflict, Iceman had boasted that he would easily defuse any DDoS attacks leveled against his site. But afterward, he Googled Master Splyntr and learned he was a world-class spammer with a botnet army. Iceman seemed loath to turn a mere critic into a full-blown enemy.

```
Don't take offense to my smartass comments. It is
true that if someone attacks me I will just track
the botnet and try to jack it or shut it down,
but it's not something I want to taunt people
with. No one needs to waste their time with such
activity, really DDoS is no fun and so don't get
the wrong idea plz. :-)
```

Mularski was beginning to see an opportunity in the upheaval gripping the underground. Nobody knew who to trust anymore; everyone was angry at everyone else. If he were to play both sides, he might make inroads against the forum administrators as they grappled for allies in the brewing battle.

He was allowed three substantive contacts. He decided to use one of them to respond to Iceman.

> No worries brotha, we're kewl. I'm a smartass myself. I got no interest in attacking. Shit, my bots aren't even configured to attack. Mailing makes me far more money! I really got no interest in doing anything that doesn't make me money, unless I have a vendetta, which I don't. And if you do get attacked, I'm also pretty good in tracking and hijacking, so hit me on ICQ 340572667 if ya need help. . . . :-) MS

Mularski watched his screen, waiting. A few minutes later, a response.

> Excellent thank you :-) BTW, do you have any suggestions for running things here, aside from the obvious organizational mess? Also, I will change it so you are a vendor and have user selectable title. (Done) I don't know if you vend mailing services with your net, but that is a cool thing to have around and I'm sure we're better off having you available for hire. Also, if you were a vendor before (or other?) then please accept my apologies for the title loss. I preserved some of the status like DM vendors, but messed up on the other forums and those didn't get preserved. Just FYI. Thanks bro :-) Also added you to VIP group.

It was a promising response. Mularski talked things over with his supervisor, then applied to headquarters for Group II authority, the lesser of two tiers of undercover engagement available to the FBI but still a step up from his previous "passive observation only" mandate. The new latitude wouldn't let him participate in crimes, but he would finally be permitted to actively engage with the underground. He named Carders Market, and everyone associated with running the site, as the investigation's targets.

The approval came quickly. But despite his encouraging words, Iceman proved a slippery target; he kept Mularski at arm's length, not confiding in him and only chatting through Carders Market's internal messaging system. The FBI agent had better luck on the other side of the battlefield. He'd been an early member of DarkMarket, and now that he was interactive, the site's founder, JiLsi, quickly identified Master Splyntr as management material. In early September, Splyntr was appointed as a moderator on the site.

The war was heating up. Despite the lessons of the August incursion, JiLsi couldn't manage to completely lock down DarkMarket. Iceman began sneaking in regularly and deleting accounts at random, just to mess with JiLsi's head. When DarkMarket retaliated with a fierce DDoS attack against Carders Market's Iranian host, Iceman fired back with a DDoS of his own against DarkMarket. Both sites groaned under the weight of the junk packets. Iceman quietly set up service at a U.S. hosting company with the bandwidth to absorb the DDoS packets, cleaning the traffic before channeling it back to his real server over an encrypted VPN.

JiLsi was tearing his hair out, voicing his frustrations to Master Splyntr. Mularski shifted his focus away from Iceman and toward the British cybercrime boss who was starting to treat him like a friend. Tentatively, he suggested that JiLsi consider turning over DarkMarket to someone seasoned in setting up bulletproof hosting. Someone accustomed to running sites that everyone hates. A spammer.

Hey, you know my background, he wrote in a chat. I'm real good at setting up servers. I secure servers all the time. I could set this up for you.

Mularski was toying with an extraordinary plan. In the past, the Secret Service and FBI had both run admins as informants: Albert Gonzalez on Shadowcrew and Dave Thomas on the Grifters. But actually running a crime forum directly would provide access to everything from the carders' IP addresses to their private communications, while giving Master Splyntr, as the site's runner, more credibility in the underground than any agent could dream of.

JiLsi expressed interest in Master Splyntr's offer, and Mularski braced himself for another trip to Washington, DC.

26

What's in Your Wallet?

```
Selling USA 100% APPROVED DUMPS
*NEW* Discounted Prices for approved dumps:
$11 MasterCard
$8 Visa Classic
$13 Visa Gold/Premium
$19 Visa Platinum
$24 Visa Signature
$24 Visa Business
$19 Visa Corporate
$24 Visa Purchasing
$19 American Express = new price drop (was 24)
$24 Discover = new price drop (was 29)
Minimum order 10 pieces.
Dumps sold by type of card. No bin list.
```

Max's hostile takeover was about fixing the community, not personal profit. But his business in stolen magstripe data was stronger than ever after the merger—he was earning a thousand dollars a day now selling dumps to carders around the world, in addition to the five to ten thousand a month he was still pulling in through his partnership with Chris.

Publicly, at FTC meetings and elsewhere, the credit card industry was doing its best to conceal the impact of the rampant magstripe theft happening worldwide. Credit leader Visa held up an industry-funded report by Javelin Strategy and Research that claimed consumers, not companies, were the source of the vast majority of identity theft and credit card fraud cases: Some 63 percent of cases originated with consumers, primarily victims of lost or stolen wallets, followed by theft by trusted associates, stolen mail, and Dumpster diving.

The report was grossly misleading, only tallying cases in which the victim knew how his information had been stolen. Visa's private numbers told the real story. Stolen wallets hadn't been the primary source of fraud since mid-2001, when credit card theft from e-commerce sites sent fraudulent "card not present" transactions—online and telephone purchases—rocketing up the chart, while every other category held steady.

In 2004, when stolen magstripe data became a massive underground commodity, losses to counterfeit cards followed the same stratospheric climb. In the first quarter of 2006, Chris Aragon–style counterfeiting edged out card-not-present fraud for the first time, topping $125 million in quarterly losses to Visa's member banks alone.

Nearly all those losses began with a price list like Max's. As Digits, Max accumulated page after page of positive reviews on Carders Market and a reputation for square dealing. It was a point of pride with Max—and a sign of the moral compartmentalization he'd practiced since childhood. Max would happily hack a carder and copy his entire hard drive, but if a customer paid him for information, Max wouldn't even consider shortchanging him.

His generosity, too, was well known. If Max had dumps that were about to expire, he'd give them away for free rather than let them go to waste. Together, his exemplary business practices and the quality of his product made Max one of the top five dumps vendors in the world, in a market traditionally dominated by Eastern European sellers.

Max was cautious with his vending. By refusing to sell dumps by BIN—bank identification number—he made it tough for the feds to identify his breaches: The government couldn't just buy twenty dumps sourced to a single financial institution and ask that bank to look for a common purchase point in its transaction records. Instead, a batch of twenty cards could belong to twenty different banks. They'd all have to cooperate with one another to nail down the source.

Additionally, only a few trusted associates knew that Digits and Iceman were one and the same: mostly admins, like Chris, a Canadian carder named NightFox, and a new recruit called Th3C0rrupted0ne.

Of everyone he'd met in the scene, it was Th3C0rrupted0ne with whom Max seemed to share the most hacking history. As a teenager, C0rrupted had discovered the warez scene on dial-up bulletin board systems, then moved into recreational hacking under the handles Acid Angel, -null-, and others. He defaced websites for fun and joined a hacking gang called Ethical Hackers Against Pedophiles—vigilante gray hats working against Internet child pornography.

Like Max, he'd once thought of himself as one of the good guys, before he became Th3C0rrupted0ne.

In other ways, they were very different. A product of a hardscrabble childhood in a big-city housing project, C0rrupted became a drug dealer at an early age and picked up his first arrest—a gun charge—in 1996 when he was eighteen years old. In college he began making fake IDs for his friends, and his online research took him to Fakeid.net, a Web bulletin board where experts like ncXVI got their start. He graduated to small check and credit card scams around the time Shadowcrew went down and then found his way to the successor sites.

Diplomatic and even-tempered, C0rrupted was universally liked in the scene and enjoyed moderator or admin privileges on most of the forums. Max promoted him to admin on Carders Market in the summer of 2005 and made him unofficial site spokesman after the hostile takeover. Max let C0rrupted in on his double identity about a week after his power play.

> So obviously I am Digits also. Might as well say it straight since I blew cover in ICQ (talking about "our forum," etc.)
>
> It is a pain in the ass trying to keep that separate from people I know and trust and like such as yourself. So there you go . . .
>
> Anyway, reasoning is, Iceman is legal. Digits is breaking the law. I assumed if I could keep it separate there would be no legal leg to stand on for coming after "me" as the forum admin.

Chris remained the greatest threat to Max's security. Every time they fought now, Max was reminded of how vulnerable he was to the only carder privy to his real-life identity. "I can't believe how much you know about me," he'd spit out, angry at himself.

Meanwhile, Chris had been trying to drive Max into pulling one big score, something that would catapult them both out of the crime business for good and maybe fund a new legitimate start-up for Chris in Orange County. He'd crafted a flowchart and a step-by-step plan for each of them to follow; he called it the "Whiz List."

Max was supposed to infiltrate banking networks and gain the power to direct millions of dollars to accounts specified by Chris. He'd delivered on his end—from the very start of their partnership, back when he was working from Chris's garage, he'd been breaching small banks and savings and loans. He was in hundreds of them now and could transfer money out of customers' accounts at will. But the scheme was hung up on Chris's end. Chris had to find a safe harbor for the money Max would steal—an offshore repository where they could park the cash without it being recalled by the victim bank. So far, he'd failed.

So when, in September, Max got his hands on a deadly new Internet Explorer zero day, he shared the news not with Chris but with a different partner, one who had more knowledge of international finance, the Carders Market admin called NightFox.

The security hole was a monster: another buffer overflow, this time in the Internet Explorer code designed to let websites draw vector graphics on a visitor's screen. Sadly for Max, Eastern European hackers had found the bug first, and they'd been using it. A computer security company had already found the Russian exploit code infecting visitors to an Internet porn site and sent it to Microsoft. The Department of Homeland Security had issued a blunt warning to Internet Explorer users: "Do not follow unsolicited links."

The word was out, but there was no patch. Every Internet Explorer user was vulnerable. Max got his copy of the Russian exploit in the early morning hours of September 26 and informed NightFox enthusiastically.

"Assume we get a free pass today to own whatever company we want," Max wrote over Carders Market's messaging system. "There you go. No limits. Visa.com. Mastercard.com. egold.com. Whatever you can get the employee e-mails for. Google. Microsoft. Doesn't matter. It's all equally ownable right now."

Microsoft pushed out a patch later that day, but Max knew that even the most secure company would take days or weeks to test and install the update. The Russian exploit was already detected by antivirus software, so he modified it to change its signature, running it through his antivirus lab to verify that it was now undetectable.

The only thing left was the social engineering: Max had to trick his targets into visiting a website loaded with the exploit code. Max decided on the domain name Financialedgenews.com, and set up hosting at ValueWeb.

NightFox came back with the target list: CitiMortage, GMAC, Experian's Lowermybills.com, Bank of America, Western Union Money-

Gram, Lending Tree, and Capital One Financial, one of the largest credit card issuers in the country. NightFox had vast databases of internal corporate e-mail addresses he'd acquired from a "competitive intelligence" firm, and he sent Max thousands of them, spread across all the targets.

On September 29, Max fired up his spamming software and flung a personalized e-mail at his victims. The message was from "Gordon Reily," with the return address g.reily@lendingnewsgroup.com.

```
I am a reporter for Lending News doing a follow
up story on the recent leak of customer records
from Capital One. I saw the name Mary Rheingold
in the article from Financial Edge and would like
to interview you for a follow up piece.

  http://financialedgenews.com/news/09/29/
Disclosure_CapitalOne

  If you have time I would greatly appreciate an
opportunity to further discuss the details of the
above article.
```

Each copy of the message was customized, so every employee would think he or she was mentioned by name in the notional Financial Edge article. At Capital One, 500 employees got the message, from executives to PR spokespeople and IT workers. About 125 of them clicked on the poisoned link and were sent to a page loaded with generic finance industry news. While they puzzled over the page, a hidden payload zipped through the corporate firewall and onto their machines.

The software opened a back door that would allow Max to slip in at his leisure and scour the victims' hard drives for sensitive data, sniff the banks' internal networks, steal passwords. It wasn't much different from what he'd done to thousands of Defense Department computers a lifetime ago. Back when it was all just fun and games.

27

Web War One

Keith Mularski stood at the podium, his PowerPoint presentation filling an LCD big-screen at his back. In front of him were fifteen senior FBI officials and Justice Department lawyers, sitting around the conference room table at Justice headquarters. They were riveted. Mularski was proposing something that had never been done before.

Group I "sensitive circumstances" authorizations were a rare thing in the bureau. Mularski first wrote out a twenty-page proposal, addressing every aspect of the plan and gathering legal opinions from FBI lawyers for each. The FBI's general counsel was excited about the possibilities; if it were approved, the operation could set a precedent for future online undercover work.

The biggest obstacle for the Justice Department's Undercover Review Committee was the third-party liability issue of letting crimes unfold over a website owned and operated by the U.S. government. How would Mularski mitigate the damage so innocent people and institutions wouldn't suffer? Mularski had an answer at the ready. The criminal activity on DarkMarket was going to take place whether the FBI ran the forum or not. But with the bureau controlling the server, and Master Splyntr leading the site, the FBI could potentially intercept large amounts of stolen data that would otherwise flow freely through the black market. His proposal stipulated that any financial data would be sent immediately to the affected banks. Stolen credit cards could be canceled before they were used.

The meeting lasted twenty minutes. When he returned to Pittsburgh on October 7, Mularski had written approval to acquire DarkMarket. Iceman was still listed as a subject of the undercover operation, but now JiLsi and DarkMarket's other leaders were the primary targets.

Once his wife went to bed, Mularski settled in front of his couch, turned on *Saturday Night Live,* and looked for JiLsi on ICQ. After some pleasantries, he got down to business. DarkMarket was under yet another DDoS attack, and Mularski, as Master Splyntr, was ready to take the site onto a secure server—JiLsi need only say the word, and his problems with Iceman would be history.

JiLsi had some reservations. DarkMarket was his baby, and he didn't want to be perceived by the community as ceding control. That wouldn't be a problem, Mularski explained. Master Splyntr would be a stealth administrator. Nobody but he and JiLsi would know he was running the site. To everyone else, he'd still just be a moderator.

"Bro," JiLsi typed back. "Get your server ready. We moving."

Mularski went to work at once. He rented a server from a Texas-based hosting company called the Planet and went to the underground to shore it up, buying $500-a-month DDoS protection services from a Russian named Quazatron and paying for it in e-gold. Quazatron configured the site so its public face was at Staminus, a DDoS-resistant high-bandwidth hosting company. The company's pipes could withstand a deluge, and Quazatron's software would channel only the legitimate traffic to DarkMarket's real server behind the scenes.

Everything would be done the way an Eastern European cybercrook would do it. When Mularski wanted to log in to the site's back end, he'd go through KIRE, a Virginia company offering Linux "shell accounts"—a service that lets IRC users connect to chat rooms without being traced to their home IP addresses. Nobody would see that the Polish spam king was logging in from Pittsburgh.

Once the move was complete, Mularski went to court and won a sealed search warrant against his own server, allowing him to riffle through DarkMarket's user database, access logs, and private messages.

There was one more thing to do. Post-Shadowcrew, it was de rigueur for carder forums to make users click on a terms-of-service agreement prohibiting illegal content and stipulating that the site's operators weren't responsible for anything on the board. Forum runners believed the legalistic language might shield them from prosecution. DarkMarket had a particularly long and detailed user agreement, so nobody noticed when Master Splyntr added a line.

"By your use of this forum you agree that the administrators may review any communication sent using this forum to ensure compliance with this policy," he wrote, "or for any other purpose."

"I think it's important to note that Iceman is a foolish wannabe hacker who goes around and hacks sites for fun and pleasure."

El Mariachi knew how to push Iceman's buttons. After the hostile takeover, Dave Thomas returned to the Life on the Road blog to browbeat his foe relentlessly, calling him "Iceboy," "Officer Ice," and "a fucking piece of shit on my shoes." He challenged Iceman to meet him in person, so they could resolve their dispute like men. And he implied he could hire a hit man to track down the carding kingpin and end his life.

Max responded with growing fury. He hadn't forgotten the hassle and expense of finding a new host after Thomas shut him down in Florida. The aggressiveness he'd kept buried since Boise boiled from his gut and into his fingertips. "You small dick limp sack of shit. I could fucking tear you apart with my bare hands but a COWARD snitch like yourself would call the cops and scramble for a weapon at the first sight of me," he wrote. "You better pray to your god that I am never outed, because not only will you look like even more of a jackass than you already do, but then I will have no inhibition about coming over and wringing your snitch punk neck."

When he calmed down, he sent Thomas a private e-mail. He'd been thinking about taking down Carders Market and retiring his Iceman identity. It wouldn't be a surrender; rather, it was the most serious threat imaginable to Thomas's campaign.

> You haven't read the Art of War, have you, cunt? You know NOTHING about me. I know EVERYTHING about you.
>
> I kill CM, I kill Iceman, then what do you have you punk bitch? Shadowboxing?? You are FUCKED. An enemy who will fuck you over constantly for years, that you have NO DEFENSE and NO TARGET for retribution.
>
> I am your worst nightmare you little bitch, and you and your family will be feeling the effects of the money you cost me for a long, long time.

Two days later, Max proved he was serious. He hacked into El Mariachi's website, the Grifters, which Thomas had turned into a semi-legitimate security site dedicated to watching the carding forums. Max wiped the hard drive. The site never came back.

Iceman announced his triumph in a final public message to the blog. "I have nothing to prove, and now having beat down David Renshaw Thomas, federal snitch, I make my exit," he wrote. "Unlike you people, I pay attention to my own business. Learn a lesson. Move on and leave me the fuck alone."

But Max wasn't going to be able to slip back into the shadows. Two reporters from *USA Today* had taken notice of the public carder war and confirmed the details of the hostile takeover with security firms watching the forums. The morning after Max declared victory over El Mariachi, delivery drivers around the country plunked down Thursday's edition of the paper on more than two million doorsteps from coast to coast. There, on the front page of the business section, was the whole sordid tale of Iceman's annexation of the carding sites.

By letting his ego lead him into a public battle with David Thomas, Max had gotten Iceman into the largest-circulation daily in America.

"The Secret Service and FBI declined to comment on Iceman or the takeovers," the article read. "Even so, the activities of this mystery figure illustrate the rising threat that cybercrime's relentless expansion—enabled in large part by the existence of forums—poses for us all."

The story wasn't a surprise; the reporters had approached Iceman for comment, and Max had e-mailed a long one, lobbing his Craigslist defense. His views didn't make it into the article, and the story only made Max more defiant. He added a quote from the piece to the top of the Carders Market login page: "It's like he created the Wal-Mart of the underground."

Max showed the article to Charity. "I seem to have created quite a stir."

Chris was apoplectic when he learned that Max had corresponded with the journalists. He'd watched as Max burned hours squabbling with Thomas. Now his partner was giving press interviews?

"You've lost your fucking mind," he said.

Max was swamped. Vouch requests were pouring into Carders Market in a torrent. The *USA Today* article seemed to bring out every street-level hood hoping to break into computer fraud. The site picked up over three hundred new members overnight. Two weeks later, they were still coming in.

He offloaded as much of the work as he could to his admins. Max had other things to worry about now. His spear-phishing attack against the financial institutions had been wildly successful, but getting past the banks' firewalls had turned out to be the easy part. Bank of America and Capital One, in particular, were huge institutions, and Max was lost in their vast networks. He could easily spend years on either one, just looking for the data and the access he needed to make a big score. Max was having trouble staying motivated for the mind-numbing follow-through to his intrusions; cracking the networks had been the fun part, and now that was over.

Instead, Max put the banks on the back burner to focus on the carding war. Max's new hosting provider was getting complaints about the rampant criminality on Carders Market. Max saw one of the e-mails, sent from an anonymous webmail account. On a hunch, Max tried logging in to the account with JiLsi's password. It worked. JiLsi was trying to get him shut down.

Max retaliated by hacking into JiLsi's account on the Russian forum Mazafaka and posting an avalanche of messages reading, simply, "I'm a fed." Then he went public with the evidence of JiLsi's malfeasance; snitching to Carders Market's hosting company was a scummy tactic.

DarkMarket just didn't have the decency to die. Max could have dropped the database again, but it would do no good—the site had come back before. His DDoS attacks had become ineffective, too. Overnight, DarkMarket had come into expensive high-bandwidth hosting and erected dedicated e-mail and database servers. It was suddenly a hard target.

Then Max heard an intriguing rumor about DarkMarket.

The story involved Silo, a Canadian hacker known for an uncanny ability to juggle dozens of false handles in the community, effortlessly switching writing styles and personalities for each one. Silo's second claim to fame was his compulsive back-dooring of other carders. He was constantly posting software with hidden code that would let him spy on his peers.

Both traits were at play when Silo registered an account at DarkMarket under a new handle and submitted a piece of hacking software for vendor review. True to form, Silo had secreted a hidden function in the software that would smuggle a user's files out to one of Silo's servers.

When Silo looked at the results, he found a small cache of blank Microsoft Word templates, including a "malware report" form. The templates carried the logo for an organization called the National Cyber Forensics and Training Alliance in Pittsburgh. Max looked them up; it was a fed shop. Someone connected with DarkMarket was working for the government.

Determined to investigate, Max breached DarkMarket again through his back door. This time, it was a reconnaissance mission. He dropped into a root shell and entered a command to bring up the recent login history and then started down the list in another window, checking the public registration records for each of the Internet IP addresses used by the administrators. When he got to Master Splyntr, he stopped. The supposedly Polish spammer had connected from an IP address belonging to a private corporation in the United States called Pembrooke Associates.

He pulled up the Whois.net registration records for the company's website, Pembetal.com. The mailing address listed was a PO box in Warrendale, Pennsylvania, twenty miles north of Pittsburgh. There was also a phone number.

Another click of his mouse, another browser window—the reverse white pages at Anywho.com. He entered the phone number and this time got a real street address: 2000 Technology Drive, Pittsburgh, Pennsylvania.

It was the address he'd already found for the National Cyber Forensics and Training Alliance. Master Splyntr was a fed.

28

Carder Court

Keith Mularski was screwed.

He got the word first from an agent at the Secret Service field office across town. "I think you may be in some trouble." One of their myriad informants heard that Iceman had uncovered incontrovertible proof that Master Splyntr was either a snitch, a corporate security spy, or a federal agent. Iceman had forged a temporary alliance with his sometime enemy Silo and was preparing a comprehensive presentation for the leadership of Carders Market and DarkMarket. Iceman and Silo were going to put Master Splyntr on trial.

It had begun with Silo's code. Master Splyntr's reputation as a spammer and programmer made him DarkMarket's go-to guy for malware reviews. It was one of the perks of his undercover operation: Mularski got the first look at the underground's latest attack code and could pass it to CERT, who would in turn give it to all the antivirus companies. The malicious code would be detectable even before it went on the black market.

This time, Mularski had assigned the code as a training exercise to one of the CMU students interning at NCFTA. As standard procedure, the student ran the program isolated in a virtual machine—a kind of software petri dish that could be scrubbed afterward. But he forgot that he had a thumb drive in the USB port. The drive was loaded with blank malware report forms containing the NCFTA logo and mission statement.

Before the intern realized what was happening, the documents were in Silo's hands.

Six DarkMarket admins and moderators had gotten a copy of Silo's code. Now the Canadian knew that one of them was a fed.

Silo was a wild card. In real life, he was Lloyd Liske, a Vancouver auto shop manager and credit card forger who'd been busted a few months after Operation Firewall. When he was sentenced to eighteen months of house arrest, Liske changed his surname from Buckell and his handle from Canucka, and reemerged in the carding scene.

Now the Canadian was untouchable. It was widely known in law enforcement circles that Silo was an informant for the Vancouver Police Department. That's why he was always back-dooring other hackers: The Trojan horse that infiltrated NCFTA wouldn't have been intended to expose a law enforcement operation; it was just Silo trying to gather intelligence on DarkMarket members for the police.

Silo had no allegiance to the FBI, but he probably wouldn't have gone out of his way to expose a bureau undercover operation. Unfortunately, Iceman had learned about the discovery and staged his reconnaissance raid on DarkMarket. That's where Mularski's own personal screwup came into play. He normally logged in to DarkMarket through his KIRE shell, hiding his location. But JiLsi was a demanding boss, constantly hitting Master Splyntr with maintenance tasks—like swapping in a new banner ad—that simply had to be performed at once. Sometimes KIRE was down when Mularski got one of these requests, and he'd take a shortcut and log in directly. Iceman had caught him.

Even then, he should have been relatively safe. The office broadband service was set up under the name of a dummy corporation, with a phone number that rang to an unanswered VoIP line in the communications room. The phone line was supposed to be unlisted. Somehow, though, it wasn't, and Iceman had gotten the address and recognized it as the NCFTA's.

Mularski walked hurriedly to the communications room, swiped his access card, keyed in the door code, and locked himself inside. He picked up the secure line to Washington. The FBI agent didn't sugarcoat his report to the brass. After all his work winning undercover authority to take over DarkMarket, getting a buy-in from senior Justice Department and bureau officials, Iceman was going to blow them out of the water just three weeks into the operation.

Max struggled with how to handle the exposé—after his attacks on DarkMarket, he knew his findings would be viewed as partisan mudslinging. He considered shuttering Carders Market before exposing Master Splyntr, to avoid the perception that the whole thing was just another volley in the carding wars. Instead, he decided to send his new lieutenant, Th3C0rrupted0ne, to represent his site.

The trial was held over Silo's "Carder IM"—a free, supposedly encrypted instant messaging program the Canadian hacker offered as an alternative to AIM and ICQ, supported by display ads for dumps vendors. Matrix001 showed up from the DarkMarket side—JiLsi was busy with the fallout from Max's attack on Mazafaka. Silo and two other Canadian carders were also present. Silo opened the meeting by handing out a compressed RAR file containing the evidence gathered by him and Iceman.

When some of the carders opened the file, their antivirus software went wild. Silo had back-doored the evidence; not a promising start to a summit meeting.

C0rrupted and Silo walked them through the case: Silo's document templates showed that someone at NCFTA held a privileged position on DarkMarket, and the access logs Iceman had stolen proved that Master Splyntr was the mole.

"One hundred percent undeniable proof," wrote C0rrupted. "We worked hard to try and make peace, and if we go public LE [law enforcement] is going to come after us HARD. But if we don't say anything, we are responsible for all those who get fucked over."

"This is for real dude," said Silo.

Matrix was unconvinced. He ran his own Whois on the Pembrooke Associates domain name and got back an anonymous listing through Domains by Proxy: no street address, no phone number. "Blah," Matrix typed. "You did not even verify the whois info and the company, did you? Who passed you that stuff?"

"That's not my stuff," wrote Silo. "That's Iceman."

"So you believe every shit which is pasted to you? Without even verifying it?"

Silo's evidence was no more convincing to Matrix: The NCFTA templates contained spelling and formatting errors—would the FBI, or a nonprofit security group, really do such shoddy work? Moreover, Iceman's contempt for DarkMarket was well-known, and Silo was a constant annoyance on the board.

The conversation grew heated. C0rrupted dropped out, and the others fell silent while Silo and Matrix began exchanging insults. "What in the whole world should make me trust you?" asked Matrix.

"Don't," Silo finally said. "Don't trust me. Get the fuck off my IM . . . Go get busted."

Mularski was excluded from the meeting, but when it concluded, Matrix sent Master Spyntr a transcript. The agent was pleased to see his last-minute cleanup had worked: As soon as he'd learned about Iceman's plans to expose him, he'd contacted the domain registrar and got the company to scrub the Pembrooke Associates name and phone number from the records. Then he asked Anywho to take out its listing for the undercover phone line. The cover-up was sure to convince Iceman all the more that Master Splyntr was a fed, but nobody else was able to independently verify his findings.

Now Mularski went into spin control over ICQ. He told Matrix and anyone else who'd listen that he was innocent. He directed the carders' attention to the logs, highlighting all the occasions he'd logged in from

KIRE's IP address. Those are my logins, he wrote. I don't know who those other logins are.

Then he spun and attacked. The doubt Iceman had sown about JiLsi worked to his advantage. Things were going crazy, he wrote. JiLsi had been acting suspiciously. For one thing, he'd instructed Master Splyntr not to tell anyone that he was running the server. And while JiLsi cultivated the impression that DarkMarket was hosted in a country out of reach of western law enforcement, he was actually hosting it in Tampa, Florida, where the feds could just waltz in any time and serve a search warrant. It was odd behavior indeed.

JiLsi protested his innocence, but it was looking bad for him. Master Splyntr publicly thanked Iceman for bringing the matter to his attention and said he'd move DarkMarket out of the United States at once.

Mularski reached out to law enforcement contacts in Ukraine, and they helped him quickly get hosting there. In the blink of an eye, DarkMarket was in Eastern Europe. Most of the carders had to agree that no fed would move a sting site to a former Soviet state.

There was no formal verdict, but a consensus formed that Master Splyntr was innocent. They weren't too sure about JiLsi.

When the controversy subsided, Mularski returned to the routine business of running his undercover operation. He was at his desk filling out reports a couple of weeks later when he got a call from another agent.

Special Agent Michael Schuler was a legend among the bureau's cybercrime agents. It was he who'd hacked into the Russians' computers in the Invita sting. Now stationed in the Richmond, Virginia, field office, Schuler was calling about a breach at nearby Capital One. The bank's security officials had detected an attack using an Internet Explorer exploit. They'd sent Schuler a copy of the code, and he wanted Mularski to get one of the NCFTA's geeks to take a look at it.

Mularski listened as Schuler described his investigation to date. He'd focused on the fake news website, Financialedgenews.com, used to deliver the malware. The domain was registered to a false identity in Georgia. But when the registrar, Go Daddy, checked its records, it found the same user had once registered another address through the company.

Cardersmarket.com.

Mularski recognized the significance at once. Iceman positioned himself as the innocent operator of a website that happened to discuss illegal activity. Now Schuler had evidence that he was also a profit-oriented hacker, one who'd broken into the network of the fifth-largest credit card issuer in America. "Dude, you got the case!" Mularski laughed. "You got the case *right now* on the guy we were just trying to target on our Group II. We've got to work together on this."

Across town, Secret Service agents at the Pittsburgh field office had made a discovery of their own about Iceman: An informant tipped them off that Carders Market's kingpin had a second identity as the dumps vendor Digits. Four days after the *USA Today* article, the agents exploited that knowledge by having a second snitch make a controlled buy from Digits: twenty-three dumps for $480 in e-gold.

It was more than they needed for a felony charge.

29

One Plat and Six Classics

Keith Mularski hadn't known what he was in for when he took over DarkMarket.

His days were crazy now. He'd start at eight in the morning, logging in to his undercover computer at the office and checking for overnight ICQ messages—any urgent business for Master Splyntr. Then he'd hit DarkMarket and make sure it was up and running. It was always hit-or-miss with Iceman on the loose.

Next came the drudgery of backing up the SQL database. Iceman had somehow dropped the tables twice since his failed attempt to expose Mularski, so now the backups were a part of Mularski's morning routine. They served an investigative function as well: While the database was being copied, a simple script authored by an NCFTA coder scanned every line for sixteen-digit numbers beginning with the numerals 3 through 6. The stolen credit card numbers would be automatically sorted by BIN and sent to the proper banks for immediate cancellation.

Next, Mularski had to review all the private messages, pick out the interesting chats, and check them into the FBI's central ELSUR electronic surveillance database. An hour or two of report writing followed. As Master Splyntr, Mularski had begun his own modest cash-out operation. Some banks had agreed to issue him disposable dumps as bait, with fake names but real lines of credit that the FBI would cover out of its investigative budget. Mularski handed them out with PINs to carders around the country, while the financial institutions reported back daily on where and when each withdrawal took place. Mularski had to pass the information to the local agents in whatever city his cashers were operating from, which meant writing up a detailed memo each time.

At three, when the carders came online in force, Mularski's second life shifted into high gear. Everyone wanted something from Master Splyntr. There were disputes to settle, like a dumps vendor complaining that his ad wasn't displayed as prominently as a competitor's, or a vendor facing accusations of ripping off a customer. Beggars approached him asking for free dumps or spamming services.

Mularski went home at the end of the day, only to log on again. Keeping his credibility as Master Splyntr meant he had to work the same hours as a real carder, so every night saw Mularski on the sofa at home, the television turned to whatever was on, his laptop open and online. He was on DarkMarket, and AIM, and ICQ, answering questions, assigning reviewers, approving vendors, and banning rippers. He stayed online and in character until two in the morning, nearly every day, dealing with the underground.

To ingratiate himself to his primary targets, he'd give them gifts or sell them discounted merchandise, supposedly purchased with stolen credit cards but actually paid for by the bureau. Cha0, a Turkish crime boss and DarkMarket admin, coveted an $800 lightweight PC sold in the States, so Mularski shipped two of them off to Cha0's drop address in Turkey. Playing Santa Claus was in his job description now: He had to appear to be running ops and making money, and he sure as hell wasn't going to spam anyone.

Being a cybercrime boss, he was discovering, was hard work.

When he traveled or vacationed, he had to let the forum know in advance—even a brief unexplained absence would invite suspicion that he'd been busted and turned. In January 2007, he let the board know that he'd be on a plane for a while. He didn't say where or why. He was going to Germany to talk with prosecutors about DarkMarket's cofounder Matrix001.

Among other things, Matrix001 was DarkMarket's resident artist par excellence. He created and sold Photoshop templates used by forgers to produce credit cards or fake ID. He had them all: Visa, MasterCard, American Express, Discover, the U.S. Social Security card, notary seals, and driver's licenses for several states. His template for an American passport sold for $45. A Bank One Visa was $125.

Matrix001 and Master Splyntr had grown tight since the attempted exposé three months earlier: Mularski and the German both liked video games, and they chatted about the latest titles well into the night. They talked business, too, and Matrix001 had confided that he received wire transfers for some of his sales in the town of Eislingen in southern Germany. That was the first clue to tracking him down.

From there, it was a matter of following the money. Like virtually all carders, Matrix preferred to be paid by e-gold, an electronic payment system created by a former Florida oncologist named Douglas Jackson in 1996. A competitor to PayPal, e-gold was the first virtual currency backed by deposits of actual gold and silver bullion held in bank vaults in London and Dubai.

It had been Jackson's dream to forge a true international monetary system independent of any government. Criminals loved it. Unlike a real bank, e-gold took no measures to verify the identity of its users—account holders included "Mickey Mouse" and "No Name." To get money in or out of e-gold, users availed themselves of any of hundreds of independent e-gold exchangers around the world, businesses that would accept bank transfers, anonymous money orders, or even cash in hand and convert it to e-gold for a cut. Exchangers took another slice when a user wanted to convert in the other direction, changing the virtual money into the local currency or receiving it by Western Union, PayPal, or wire transfer. One company even offered a preloaded ATM card—the "G-Card"—that would let account holders withdraw their e-gold from any cash machine.

By all evidence, criminals were e-gold's bread and butter. By December 2005, the company's internal investigations had identified more than three thousand accounts involved in carding, another three thousand used for buying and selling child porn, and thirteen thousand accounts linked to various investment scams. They were easy enough to spot: the "memo" field in child porn transactions would read, for example, "Lolita"; in Ponzi schemes, "HYIP," for "high-yield investment program." Carders included shorthand descriptions of what they were buying: "For 3 IDs"; "for dumps"; "10 classics"; "Fame's dumps"; "10 M/C"; "one plat and six classics"; "20 vclassics"; "18 ssns"; "10 AZIDs"; "4 v classics"; "four cvv2s"; "for 150 classics."

For a long time, e-gold largely turned a blind eye to the criminal trade; employees locked down some accounts used by child porn sellers but didn't stop them from transferring out their money. But the company's attitude changed dramatically in December 2005, when FBI and Secret Service agents executed a search warrant at e-gold's Melbourne, Florida, offices and accused Jackson of running an unlicensed money transfer service.

Jackson began voluntarily searching his database for signs of criminality and sending tips to the only agency that wasn't trying to put him in jail, the U.S. Postal Inspection Service. His newfound commitment to law and order was a boon to Mularski. Through Greg Crabb and his team at the post office, Mularski asked Jackson for information about Matrix001's e-gold account, which was under the alias "Ling Ching." When Jackson looked in his database, he found that the account had originally been set up under another name: Markus Kellerer, with a street address in Eislingen. In November, Mularski sent a formal request for assistance to the German national police through the U.S. consulate in Frankfurt. The police confirmed that Kellerer was a real person and not just another alias, and Mularski booked his flight to Stuttgart.

Matrix001 would be the first arrest from the DarkMarket sting. Mularski would have to find someone else to chat with about video games.

. . .

Once he was back in Pittsburgh, Mularski began working a new, far-fetched theory about Iceman. He'd been running down every "Iceman" he could find—there'd been an Iceman on Shadowcrew and others on IRC. They always turned out to be red herrings. Now Mularski was toying with the idea that his Iceman didn't really exist.

It was Iceman's supposed collaboration with the Canadian informant Lloyd "Silo" Liske that intrigued him. Silo had worked with Iceman to try to expose Mularski. That, in itself, didn't mean much—informants often call out suspected cops and snitches to deflect suspicion from themselves. But Silo had told his handler at the Vancouver Police Department that he'd hacked Iceman's computer, yet when push came to shove, he couldn't produce Iceman's real name or even a good Internet IP address. And it turned out that Silo had dozens of e-gold accounts—one of them under the name "Keyser Söze."

If Liske was a fan of *The Usual Suspects,* it might occur to him to create a phantom criminal mastermind and then feed law enforcement false information about the supposed kingpin in his role as an informant.

Mularski flew to Washington and presented his theory to the Secret Service at their headquarters. It was shot down at once. They were working closely with Silo's handler at the Vancouver Police Department, and they knew Silo as one of the good guys.

The Secret Service had run down some false leads themselves. In a lab in the Pittsburgh field office, the agents had a whiteboard scrawled with handles and names connected by squiggles and lines. Many of the names were crossed out. It was their ever-changing road map to Iceman and his world.

Mularski returned to Pittsburgh, and both agencies resumed their search for the real Keyser Söze of the cyberworld—the elusive hacking kingpin Iceman.

30

Maksik

Max could see what was coming. With an FBI agent at the helm, DarkMarket was going to put a lot of carders in prison. But like Cassandra from Greek mythology, he was cursed to know the future and have nobody believe him.

Between the *USA Today* article and his failed attempt to expose Master Splyntr, Max could feel the heat coming at him. In November, he declared Iceman's retirement and made a show of handing control of the site to Th3C0rrupted0ne. He secluded himself while things cooled down and three weeks later took back the board under another handle. Iceman was dead; long live "Aphex."

Max was getting tired of the tight quarters at the Post Street Towers, so Chris brought Nancy, one of his cashers, up to San Francisco to rent Max a one-bedroom at Archstone's towering Fox Plaza corporate apartment complex in the financial district. She posed as a sales representative at Capital Solutions, a corporate front Aragon used to launder some of his income. Tea, back from her trip to Mongolia, was conscripted to sit in the apartment and accept delivery of a bed, paid for with her legitimate American Express card. Chris reimbursed her afterward.

By January 2007, Max was back in business at his new safe house, with a stew of Wi-Fi brewing outside. Fox Plaza was a giant step up in luxury from the Post Street Towers, but Max could afford it—he could pay a month's rent with a couple of good days of dumps vending. As Digits, Max was now regarded by some carders as the second-most-successful magstripe vendor in the world.

The number one spot was firmly occupied by a Ukrainian known as Maksik. Maksik operated outside the carding forums, running his own Web-based dispensary for his stolen cards at Maksik.cc. Buyers would begin by sending Maksik upfront money by e-gold, WebMoney, wire transfer, or Western Union. That would buy them access to his website, where they could select the dumps they wanted by BIN and type of card and place an order. On his end, Maksik would press a button to approve the transaction, and the buyer would get an e-mail with the dumps he'd ordered, straight from Maksik's massive database of stolen cards.

Maksik's wares were phenomenal, with a high success rate at the register and a mammoth selection of BINs. Like Max's, Maksik's cards came from swipes at point-of-sale terminals. But instead of targeting scores of small stores and restaurants, Maksik got his cards from a smaller number of giant targets: Polo Ralph Lauren in 2004; Office Max in 2005. In three months, Discount Shoe Warehouse lost 1.4 million cards taken from 108 stores in 25 states—straight into Maksik's database. In July 2005, a record-breaking 45.6 million dumps were stolen from the TJX-owned retail chains T. J. Maxx, Marshalls, and HomeGoods.

There was a time when such breaches might have remained a secret between the hackers, the companies, and federal law enforcement—with the victim consumers kept in the dark. To encourage companies to report breaches, some FBI agents had an unofficial policy of keeping company names out of indictments and press releases, protecting corporations from bad publicity over their shoddy security. In the 1997 Carlos Salgado Jr. case—the first large-scale online credit card heist—the government persuaded the sentencing judge to permanently seal the court transcripts, for fear the targeted company would suffer "loss of business due to the perception by others that computer systems may be vulnerable." Consequently, the eighty thousand victims were never notified that their names, addresses, and credit card numbers had been offered for sale on IRC.

In 2003, the state of California effectively ended such cover-ups when the legislature enacted SB1386, the nation's first compulsory breach-disclosure law. The law requires hacked organizations doing business in the Golden State to promptly warn potential identity theft victims of a breach. In the years that followed, forty-five other states passed similar legislation. Now no significant breach of consumer data remains a secret for long, once detected by the company and the banks.

The headlines over the giant retail breaches only added luster to Maksik's product—he didn't try to hide the fact that he was vending the dumps from the retail chains. When the TJX attack made news in January 2007, the details that emerged also confirmed what many carders already suspected: the Ukrainian had a stateside hacker supplying him with dumps. Maksik was a middleman for a mystery hacker in the States.

In mid-2006, the hacker was apparently in Miami, where he parked at two TJX-owned Marshalls outlets and cracked the stores' Wi-Fi encryption. From there, he hopped on the local network and swam upstream to the corporate headquarters, where he launched a packet sniffer to capture credit card transactions live from the Marshalls, T. J. Maxx, and HomeGoods stores around the country. The sniffer, an investigation would later find, ran undetected for seven months.

Max had a rival in America, and a damn good one.

Thanks in large part to Maksik's hacker and Max Vision, the popular consumer impression that Web transactions were less secure than real-life purchases was now completely false. In 2007, the majority of compromised cards were stolen from brick-and-mortar retailers and restaurants. The large retail intrusions were compromising millions of cards at a time, but breaches at smaller merchants were far more common—Visa's analysis found 83 percent of credit card breaches were at merchants processing one million Visa transactions or less annually, with the majority of thefts taking place at restaurants.

Max tried to keep the sources of his dumps a secret, falsely claiming in his forum posts that the data came from credit card processing centers to throw investigators off track. But Visa knew that restaurant point-of-sale terminals were being hit hard. In November 2006, the company issued a bulletin to the food service industry warning about hack attacks unfolding through VNC and other remote-access software. Max, though, continued to find a steady stream of vulnerable eateries.

But for Max, it wasn't enough. He hadn't gone into the data-theft business to be second-best. Maksik was costing him money. Even Chris was now buying from both Maksik and Max, going with whichever vendor offered him a good deal on the best dumps.

At Max's direction, Tea befriended the Ukrainian over the course of months and urged him to start vending on Carders Market. Maksik declined graciously and suggested she visit him sometime in Ukraine. Rebuffed, Max took the gloves off and got Tea to send Maksik a Trojan horse program, hoping to get control of the Ukranian's database of dumps. Maksik laughed off the hacking attempt.

If he'd known, Max might have taken comfort in the fact that he wasn't the only one frustrated by Maksik's tight security.

Federal law enforcement had been tracking Maksik since his rise to infamy in the wake of Operation Firewall. An undercover Secret Service agent had been buying dumps from him. Postal Inspector Greg Crabb had worked with law enforcement in Europe to bust carders who'd done business with Maksik, and he shared the resulting information with the Ukrainian national police. In early 2006, the Ukranians finally identified Maksik as one Maksym Yastremski, from Kharkov. But they didn't have enough evidence to make an arrest.

The United States refocused on identifying Maksik's hacking source. E-gold once again provided the entry point. The Secret Service analyzed

Maksik's accounts in the e-gold database and found that between February and May 2006, Maksik had transferred $410,750 out of his account to "Segvec," a Mazafaka dumps vendor generally thought to be in Eastern Europe. An outward transfer implied Segvec wasn't one of Maksik's customers but a supplier getting his cut.

The feds got a chance at more direct information in June 2006, when Maksik was vacationing in Dubai. Secret Service agents from San Diego worked with local police to execute a "sneak-and-peek" in his room, where they secretly copied his hard drive for analysis. But it was a dead end. The sensitive material on the drive was all encrypted with a program called Pretty Good Privacy. It was good enough to stop the Secret Service in its tracks.

Carders like Maksik and Max were at the fore in embracing one of the unheralded gifts of the computer revolution: cryptography software so strong that, in theory, even the NSA couldn't crack it.

In the 1990s the Justice Department and Louis Freeh's FBI had tried hard to make such encryption illegal in the United States, fearing that it would be embraced by organized crime, pedophiles, terrorists, and hackers. It was a doomed effort. American mathematicians had decades before developed and published high-security encryption algorithms that rivaled the government's own classified systems; the genie was out of the bottle. In 1991, a U.S. programmer and activist named Phil Zimmerman had released the free software Pretty Good Privacy, which was available on the Web.

But that didn't stop law enforcement and intelligence officials from trying. In 1993, the Clinton administration began producing the so-called Clipper Chip, an NSA-developed encryption chip intended for use in computers and telephones and designed with a "key recovery" feature that would allow the government to crack the crypto on demand, with the proper legal authority. The chip was a dismal failure in the marketplace, and the project was dead by 1996.

Then lawmakers began swinging the opposite direction, talking about repealing Cold War–era export regulations that classified strong encryption as a "munition" generally prohibited from export. The regulations were forcing technology companies to keep strong crypto out of key Internet software, weakening online security; meanwhile, overseas companies weren't bound by the laws and were in position to overtake America in the encryption market.

The feds responded with a draconian counterproposal that would have made it a five-year felony to sell any encryption software in America that lacked a back door for law enforcement and government spies. In testimony to a House subcommittee in 1997, a Justice Department lawyer warned that hackers would be a prime customer of legal encryption and used the Carlos Salgado bust to illustrate his point. Salgado had encrypted the CD-ROM containing the eighty thousand stolen credit card numbers. The FBI had only been able to access it because the hacker gave his supposed buyer the key.

"We were lucky in this case, because Salgado's purchaser was cooperating with the FBI," the official testified. "But if we had discovered this case another way, law enforcement could not have penetrated the information on Salgado's CD-ROM. Crimes like this one have serious implications for law enforcement's ability to protect commercial data as well as personal privacy."

But the feds lost the crypto wars, and by 2005 unbreakable crypto was widely available to anyone who wanted it. The predictions of doom had largely failed to materialize; most criminals weren't tech-savvy enough to adopt encryption.

Max, though, was. If all his tradecraft failed and the feds crashed through his safe house door, they'd find everything he accumulated in his crimes, from credit card numbers to hacking code, scrambled with an Israeli-made encryption program called DriveCrypt—1,344-bit military-grade crypto he'd purchased for about $60.

The government would arrest him anyway, he expected, and demand his passphrase. He would claim to have forgotten it. A federal judge somewhere would order him to disclose the secret key, and he'd refuse. He'd be held on contempt charges for maybe a year and then be released. Without his files, the government wouldn't have any evidence of his real crimes.

Nothing had been left to chance—Max was certain. He was untouchable.

31

The Trial

Jonathan Giannone, the Long Island carder Max and Chris had discovered as a teenager, was keeping a secret from everyone.

The same day Max had absorbed his competitors, Secret Service agents had arrested Giannone at his parents' house for selling some of Max's dumps to Brett Johnson, the Secret Service informant known as Gollumfun. Giannone was released on bail, but he told nobody about the bust. To him, it was just a bump in the road—how much trouble could he really get in for selling twenty-nine dumps?

The impression that he was facing a slap on the wrist was bolstered when the judge in South Carolina lifted his travel restrictions a month after his arrest. Giannone promptly flew into Oakland Airport on a carding run, and Tea picked him up and showed him around. They drove up and down the Pacific Coast Highway, and she bought him a pizza at Fat Slice on Berkeley's Telegraph Avenue. She'd always found Giannone amusing—a boastful, curly-haired white kid with hip-hop sensibilities who'd once bragged that he'd beat up a member of the New York Jets at a local bar. Now, though, they had something in common: Chris had stopped talking to Giannone around the time of his arrest, while Tea, for her part, had been ordered to return to the Bay Area so she couldn't make any more trouble with Chris's relationships. Chris had exiled them both.

Chris called Tea while they were hanging out and was surprised to hear that Giannone was in town. He had her put Giannone on the phone. "So, you take my girls out to party now?" he demanded, angry that Giannone was forging a relationship with one of his people—perhaps courting her for a cashing crew of his own.

"No, I just happen to be here and I looked her up," Giannone said a little defensively.

It would be Chris's and Giannone's last phone conversation. Giannone flew home. He kept in touch with Tea, and a few months later, he warned her that he might not be a good person to be associating with. He was pretty sure he'd been followed on his trip to the Bay Area.

"I got some heat on me right now," Giannone said.

"What kind of heat?" Tea asked. Giannone liked to affect an air of danger.

"I go to trial next week."

Federal criminal trials are rare. Faced with the long prison terms recommended by rigid sentencing guidelines, most defendants opt to take a plea deal in exchange for a slightly shortened sentence or limit their exposure by becoming an informant. Some 87 percent of prosecutions were resolved in this manner in 2006, the year of Giannone's trial. In another 9 percent of the cases, charges were dismissed before reaching a trial, the government preferring to drop a marginal case rather than risk a loss. Once a jury is seated, a defendant's chances for acquittal are about one in ten.

But Giannone liked his odds. Most cases don't hinge on the undercover work performed by an active computer criminal. Soon after he'd snitched on Giannone, Brett "Gollumfun" Johnson had gone on a four-month cross-country crime spree, pulling his IRS scam in Texas, Arizona, New Mexico, Las Vegas, California, and Florida, where he was finally nabbed in Orlando with nearly $200,000 stuffed in backpacks in his bedroom. He wouldn't make a very good witness for the prosecution.

The bailiff passed out pads and pencils to the twelve jurors, and the prosecutor began his opening statement, adopting a down-home, country tone.

"I love the Internet," he said. "The Internet is a fascinating thing. It's a place where we can entertain ourselves; we can get information; we can watch videos; we can play games; we can buy things. eBay is a great place, you can bid on things. If you can think about it, you can buy it on eBay.

"But, ladies and gentlemen, there's a side of the Internet that we don't like to think about. There's kind of a dark underbelly to the Internet, one where not trinkets or bobbles are bought, sold, and traded. There's a part of the Internet where people's *lives* are bought, sold, and traded. . . .

"You are going to see that side of the Internet. And I suspect that you are never going to look at the Internet exactly the same way again."

The trial lasted three days. The prosecutor disposed of Brett Johnson right out of the gate, acknowledging that Gollumfun was a liar and a thief who'd betrayed the trust of his Secret Service handlers. That was why the government wasn't calling him to testify. The prosecution's "star witness" would be the computer logs of Giannone's chats with the informant. The record would speak for itself.

Giannone's lawyer did his best to attack the logs. "Machines make mistakes." He argued that because the stolen credit cards were never fraudulently used, there were no victims. He reminded the jurors that nobody died or suffered physical harm.

After one day of deliberation, the verdict came in: guilty. The first federal trial of the carding underground was over. The judge ordered Giannone taken into custody.

A week later, Giannone was summoned from his cell at the Lexington County Jail. He instantly recognized the Secret Service agents waiting by the sally port, two steel doors away from freedom; the two men had been Johnson's handlers, and they'd testified at Giannone's trial.

"We want to know who this guy Iceman is," one of them said.

"Who's Iceman?" Giannone answered innocently.

The situation was serious, the agents said; they'd learned that Iceman had threatened to kill the president. Giannone asked for his lawyer, and the agents phoned him on the spot. The attorney consented to an interview in the hope of winning leniency for his client at sentencing.

In a series of meetings over the next three weeks, the agents pulled Giannone out of jail again and again, shuttling him to the same field office where Gollumfun had orchestrated his downfall. Unlike most carders, Giannone had held his mud at his arrest and taken a chance on a trial instead of cutting a snitch deal. But now he was looking down the barrel of a five-year sentence. He was only twenty-one years old.

Giannone told them everything he knew: Iceman lived in San Francisco, did a brisk business in dumps, sometimes used the aliases Digits and Generous to sell his goods. He used hacked Wi-Fi to cover his tracks. A Mongolian woman called Tea was his Russian translator.

Most crucially, he had a partner named Christopher Aragon in Orange County, California. You want Iceman? Get Chris Aragon.

The revelations electrified the agents tracking Iceman. When Keith Mularski typed Chris Aragon's name into the FBI's case management system, he found Werner Janer's 2006 proffer sessions, in which he'd named Chris's dumps supplier as a tall, ponytailed man he knew as "Max the Hacker." It got better. Way back in December 2005, Jeff Norminton had been arrested for receiving Janer's wire transfer on behalf of Aragon. He'd told the FBI about introducing Aragon to the superhacker Max Butler after his release from Taft. The interviewing agent was only interested in real estate fraud and hadn't pursued the lead.

Now Mularski and his Secret Service counterparts had a name. Giannone's statements confirmed it. Iceman had told Giannone that he was once raided as a suspect in the Half-Life 2 source-code theft. Mularski ran another search and saw there were only two U.S. search warrants executed in that investigation: one against Chris Toshok, and one against Max Ray Butler.

Iceman's identity had been hidden in the government's computers all along. Giannone had given them the password to unlock it.

Knowing Iceman's identity wasn't the same as proving it, though. The feds had enough for a search warrant, but they didn't have the location of Max's safe house. Worse, Giannone had tipped them that Iceman used DriveCrypt. That meant that even if they tracked down Max's address, they couldn't count on finding evidence on his hard drive. They could bust down Max's door, then watch him walk out of a courtroom twenty-four hours later on bail or a signature bond. With an international network of fake ID vendors and identity thieves at his beck and call, Max might vanish, never to be seen again.

They needed to sew up the case before making a move. Mularski decided Chris Aragon was the key. Thanks to Norminton, they knew all about the wire transfer and real estate fraud scheme he'd profited from almost five years earlier. If they could nail Aragon for that, they could press him to cooperate against Max.

Unaware of the net tightening around him, Max continued his round-the-clock management of Carders Market as "Aphex." Not that his new identity was really fooling anyone. He couldn't resist carrying Iceman's campaign against DarkMarket's leaders into his new persona, calling them "idiots and incompetents" and circulating the evidence he'd gathered against Master Splyntr. He was astonished that so many people didn't believe him. "DarkMarket is founded and run by NCFTA/FBI for Christ sake!"

Th3C0rrupted0ne believed Max and gave up his status on DarkMarket to work as a full-time admin on Max's board—he was devoting fourteen hours a day to the site now. But Max didn't trust him either. It was well-known that C0rrupted lived in Pittsburgh, the home of the NCFTA.

Max had developed a new test function for possible informants, and in March he'd tried it out on the carder, announcing out of the blue that he was working with a terrorist cell "and we should have a shot at killing President Bush this coming weekend." If C0rrupted was a fed, he'd be obliged to discourage the notional assassination plot, Max figured, or he'd ask for more details.

C0rrupted's response briefly assuaged Max's doubts. "Good luck with the president thing. Make sure you get the vice president as well. He is no better."

There was a lot of work to do on the board. Carders Market was hopping, with over a dozen specialized vendors: DataCorporation, Bolor, Tsar Boris, Perl, and RevenantShadow sold credit card numbers with CVV2s, stolen variously from the United States, UK, and Canada; Yevin vended California driver's licenses; Notepad would check the validity of dumps for a small fee; Snake Solid moved U.S. and Canadian dumps; Voroshilov offered identity thieves a service that could obtain a victim's Social Security number and date of birth; DelusionNFX vended hacked online banking logins; Illusionist was Carders Market's answer to JiLsi, selling novelty templates and credit card images; Imagine competed with EasyLivin' in the plastics trade.

Max tried to run a tight ship—a "military base," one carder critic groused. As in his white-hat days, he prized intellectual honesty, refusing to grant special favor to even his closest allies.

In April, C0rrupted prepared a review of Chris's latest generation of "novelty" IDs and plastics. He found them wanting—for one thing, the signature strips were printed right on the cards; you had to sign them with a felt-tip. He thought the products were worth five stars out of ten, but he asked Max if he should fluff his findings a little. "I know you and Easylivin' are close, so I wanted to know if I should post a true opinion review about these things that I felt, or if I should not be so harsh?"

"I think definitely post the truth, and if possible back it up with pics etc.," Max wrote back. "I am tight with Easylivin', but I think the truth is more important. Besides, if he is covered for, and continues to ship poor quality (damn . . . it's really that bad?) then it will reflect badly on you and Carders Market."

A bad review would cost Chris money. But Max didn't hesitate when it came to the integrity of his crime site.

32

The Mall

Chris pulled his Tahoe into the garage at Fashion Island Mall in Newport Beach, parked, and got out with his new partner, twenty-three-year-old Guy Shitrit. They walked toward the Bloomingdale's, fake American Express cards in their wallets.

Originally from Israel, Shitrit was a handsome guitar player and ladies' man whom Chris had met on Carders Market. Shitrit had been running a skimming operation in Miami, recruiting professional strippers at work and equipping them with exceedingly small skimming devices to steal patrons' magstripe data. When the strip-club managers found out, Shitrit had to get out of town in a hurry. He'd landed in Orange County, where Chris hooked him up with a fake ID, a rental car, and an apartment at the Archstone. Then they hit the stores.

Chris was close now, so close, to getting out. His wife, Clara, had brought in $780,000 on eBay in a little over three years: 2,609 Coach bags, iPods, Michele watches, and Juicy Couture clothes. She had an employee working twenty hours a week just shipping the ill-gotten merchandise. Chris added to the take with his sales of plastics and novelties on Carders Market, an enterprise that wasn't helped by Th3C0rrupted0ne's nitpicking review.

Max, he felt, was ignoring the Whiz List, their blueprint for building one big score and getting out. Chris had finally figured it out: Max didn't want to quit. He liked hacking; it was all he wanted to do. So screw him. Chris had his own exit strategy in place. He'd poured his profits into an enterprise for Clara, a denim fashion company called Trendsetter USA that already employed several full-time workers at a bright, pleasant office in Aliso Viejo. Eventually, he was certain, it would be profitable. And 100 percent legit.

Until then, he'd be busy.

Shitrit was a clotheshorse, and they'd already squandered some of their stolen credit on men's clothing for him. On this visit, they'd stay focused. They walked into the air-conditioned coolness of the Bloomingdale's and made a beeline for Ladies' Handbags. The Coach purses rested on small shelves along one wall, individually spotlit like museum exhibits. Chris and Guy each picked some out and went to the register. After some swipes at the point-of-sale terminal, they were headed for the door with $13,000 worth of Coach in their hands.

Chris was breaking his own rules by going in-store himself, but his crew was suddenly thinning. Nancy, who'd helped set up Max's new safe house, had since moved to Atlanta and was doing only a little cashing there. Liz was becoming paranoid—she was constantly accusing Chris of ripping her off, conveying her displeasure in meticulous, hand-drawn spreadsheets summing up how much Chris owed her for each in-store appearance: $1,918 from a trip to Vegas; $674 for iPods and GPS systems; $525 for four Coach purses worth $1,750. The "amount paid to me" column was zeroes all the way down. In the meantime, his newest recruit, Sarah, was balking at big-ticket items, though she was still useful for running errands. On Valentine's Day she bought Chris's presents for his wife and his girlfriend.

With the demands of vending, starting a legitimate business, and trying to resuscitate his crew, Chris now found it more efficient to pay someone else to make his plastic. He'd met Federico Vigo at UBuyWeRush. Vigo was looking for a way to pay down a $100,000 debt to the Mexican Mafia, after accepting that amount in front money to import a pallet of ephedra from China, only to have the product intercepted at the border. Chris put him to work. The counterfeiting gear was moved from the Tea House to Vigo's office in Northridge, and one of Chris's gophers was running out to the Valley a couple of times a week to collect the latest batch of credit cards hot off the presses, paying Vigo $10 for each card.

Chris and Guy left the Bloomingdale's and kept their unhurried pace back to the SUV. Chris popped the back and found a place for the new purchases amid a dozen plain brown department store bags already jostling for space, each filled with purses, watches, and a smattering of men's clothing. He closed up; they got into the car and started planning their next stop.

They were still planning when a white police cruiser zoomed into the garage. It stopped near them and disgorged two uniformed Newport Beach Police Department officers.

Chris's heart sank. Another bust.

The police booked Chris at the Newport Beach Police Station just down the road from the mall, then searched his car, turning up seventy credit cards and small amounts of Ecstasy and Xanax. Once he was fingerprinted, Chris was ushered into an interrogation room, where Detective Bob Watts handed him a Miranda waiver.

Chris signed and launched into the same basic story that had gotten him out of serious trouble in San Francisco a few years earlier. He promptly admitted his real name and confessed with evident shame to using counterfeit credit cards at Bloomingdale's and elsewhere. It was the economy, he said. He'd worked in the mortgage industry and was hit hard when the real estate market collapsed. That's when the head of an Orange County carding ring recruited him to card merchandise for a small percentage of the profits. He was just a mule.

It was a familiar tale to Watts, who'd busted low-level cashers before. It even explained Aragon's amateurish Bloomingdale's run—gobbling up thousands of dollars' worth of Coach bags at once. Bloomingdale's security people didn't like to upset the store's customers, so when they had a suspicious one, they normally called Watts or his partner, who'd arrange for a discreet traffic stop on a "vehicle code violation" to check out the suspect away from the store. If the shopper was innocent, they'd never know that Bloomingdale's had called the cops on them. Chris's and Shitrit's behavior, though, was so blatant that the store had no worries that they might be innocent. The security team called the police dispatch desk directly to make sure the men didn't get out of the parking lot.

But Watts wasn't buying Chris Aragon's hard-luck story. He'd been a detective for only eight months but a cop for seven years; the first thing he'd done when Aragon came in was run him through NCIC. He'd seen that Chris's criminal record stretched back to the seventies, and technically, he was still on probation from his most recent bust in San Francisco—for credit card fraud.

He figured he had a ringleader in his holding cell. He got a search warrant in a hurry and converged with a team of detectives and uniformed cops at the only address he could find for Chris: Trendsetter USA. One look at the baffled faces of the employees as the cops stormed the door told Watts they were innocent. After some questioning, one of the workers mentioned that their boss, Clara, ran an eBay business in the back office.

Watts opened the storage cabinets in back and took inventory: thirty-one Coach bags, twelve new Canon PowerShot digital cameras, several TomTom GPS navigators, Chanel sunglasses, Palm organizers, and iPods, all new in the box.

Clara walked into the office in the middle of the search and was promptly arrested. In her purse, Watts found several utility bills for an address in Capistrano Beach, all in different names. Clara reluctantly admitted she lived there; her face fell when Watts told her it was his next stop.

With Clara's house keys and a new search warrant in hand, the detectives arrived at the Aragon home and began their search. In Chris's home office, they found an unlocked safe in the closet. Inside were two plastic index-card cases crammed with counterfeit cards. There were more cards in the bedroom, bundled in rubber bands and stashed in the night table. An MSR206 rested on a shelf in the family room, and in the connecting garage, a box of purses sat on the floor next to the fitness machine.

Aside from the dining room and bathrooms, the only space in the house clean of evidence was the boys' comfortable bedroom. Just two twin beds, side by side, some stuffed animals and toys.

For all his talk about credit card fraud as a victimless crime, Chris had overlooked the two most vulnerable victims of his conduct. They were four and seven, and their dad wasn't coming home.

33

Exit Strategy

"That's a fed," Max said, indicating a sedan passing them on the street. Charity glanced skeptically at the Ford. American-made cars were just one of the many things that alarmed Max these days.

Weeks had passed since Chris's arrest, and reading the press coverage from Orange County, Max couldn't get over how much evidence the police had found in Aragon's home. Using Chris's payout sheets as a road map, the cops had rounded up his entire cashing crew; even Marcus, Chris's pot grower and errand boy, was busted with a hydroponic dope farm growing in his Archstone apartment. After two weeks of hunting, the police converged on Chris's credit card factory at Federico Vigo's office in the Valley, arrested Vigo, and seized the counterfeiting gear. Chris was being held on a million dollars' bail.

The entire operation had been dismantled piece by piece. They were calling it perhaps the largest identity-theft ring in Orange County's history.

"Shit, I wonder what kind of records he kept on all that," Max later wrote The3C0rrupted0ne. "I mean, if he was sloppy enough to have equipment at his house."

Max had already ditched his prepaid cell phone and instituted a "security ban" on his former partner's Carders Market account. They were routine precautions—he was largely unconcerned about the bust at first; it was, after all, just a state case. Chris had been caught red-handed at the W, too, and that time he walked away with probation.

But as the weeks passed with Chris still in jail, Max started to worry. He was noticing strange cars parked on his street—an animal control van aroused his suspicion so much he got out a flashlight to peer in the windows. Then a San Francisco FBI agent called him out of the blue to inquire about Max's long-dead arachNIDS database. Max decided to invest in a rope ladder; he kept it by the back window of the apartment he shared with Charity, in case he had to get out fast.

He'd pause every now and then to reflect on his freedom—here he was, enjoying life, hacking, while at that very moment Chris was in a jail cell in Orange County.

Max picked a random San Francisco criminal defense attorney from the yellow pages, walked into his office, and handed over a pile of cash; he wanted the lawyer to travel to Southern California to check on Chris and see if there was anything he could do. The attorney said he'd look into it, but Max never heard back from him.

It was then that Max finally learned about Giannone's bust from a news article about Brett Johnson's life as an informant. Max had lost track of Giannone, and for all his hacking, Max had never thought to run the names of his associates through the public federal court website. The news that Giannone had lost a criminal trial worried him.

"Of all the rat snitch piece of shit motherfuckers out there, Giannone is the closest to being able to finger me for the feds," he confided in a post to the private administrators' forum on Carders Market. "The little dipshit might actually be able to get the feds close to me."

Max uprooted from Fox Plaza, hiding his equipment at home until he was set up with a new sanctuary. On June 7, he picked up the keys at the Oakwood Geary, another corporate apartment building carved out of gleaming marble in the Tenderloin. He was "Daniel Chance" now, just another displaced software drone relocating to the Bay Area. The real

Chance was fifty years old and bearded, while Max was clean shaven with long hair—but the fake driver's license and genuine money order were enough to get him in.

The next evening, Max checked out a red Mustang from his neighborhood Zipcar and packed it with his computer gear. For all his paranoia, he didn't notice the Secret Service agents tailing him on the drive to the Oakwood and watching from the street as he moved into his new safe house.

A month later, Max jolted awake, shot upright in bed, and blinked into the darkness of the flat. It was just Charity; she had crawled into bed next to him, trying in vain not to wake him. He was growing jumpier every day.

"Sweetie, you can't keep doing this," Charity murmured. "You may not realize it, but I realize it. I can see it. You're getting too sucked into it mentally. You're losing focus of who you are and what you're doing."

"You're right," he said. "I'm done."

A lot of time had passed since his last prison term, he thought. Maybe he could find honest work again. NightFox had already offered him a legitimate job in Canada, but he'd turned it down. He couldn't bring himself to leave Charity. He'd been contemplating marriage, playing with the idea of luring her to Las Vegas on a vacation and popping the question there. She was fiercely independent, but she couldn't argue that he hadn't given her space.

It was time, he decided, for Max Vision, white hat, to return. It would be official. He visited the San Francisco courthouse and filled out the necessary paperwork. On August 14, a judge approved his legal name change from Max Butler to Max Ray Vision.

He already had an idea for a new website that could catapult him back into the white-hat scene: a system for disclosing and managing zero-day vulnerabilities. He could seed it with the security holes he was privy to in the underground, bringing the exploits into the white-hat world like a defector crossing Checkpoint Charlie with a suitcase full of state secrets.

But after all his work making Carders Market the top crime forum in the English-speaking world, he couldn't bring himself to just abandon it.

Max returned to his safe house. It was August, and the heat was back—the temperature topped 90 degrees outside, and higher in his studio. His CPU was threatening to burn itself alive. He turned on his fans, sat at his keyboard, and began the work of phasing out his Digits and Aphex identities.

He logged on to Carders Market and, as Digits, posted a note that he was shunting his dumps vending to Unauthorized, one of his admins. Then, as Aphex, he announced that he was retiring from carding and was selling Carders Market. He let the announcement sit for a few minutes and then took down the site. When he brought it back up, Achilous, one of his administrators in Canada, was in charge. Max created a new, generic handle for himself, "Admin," to help Carders Market's new kingpin during the transition.

He was still working on his exit strategy when an instant message popped up on his screen. It was from Silo, the Canadian carder who was always trying, and failing, to hack him. Max had tracked him down and identified him as Lloyd Liske in British Columbia. He suspected Liske was an informant.

The note was odd, a long sentence about newbies making dumb mistakes. But Silo had hidden a second message within it by strategically capitalizing nine of the letters.

They spelled out "MAXVISION."

A guess, Max thought. *Silo couldn't possibly know anything.*

It was just a guess.

. . .

The day after Max announced his retirement, Secret Service agent Melissa McKenzie and a federal prosecutor from Pittsburgh flew to California to tie up some loose ends.

The investigation was nearly complete. The Secret Service had gotten ahold of Digits's e-mail from a contact at the Vancouver Police Department—Silo's handler. Max had been using a Canadian-based webmail provider called Hushmail that provides high-security encryption, using a Java applet that decrypts a customer's messages right on his own PC instead of the company's server. In theory, the arrangement ensures that even Hushmail can't get at a customer's secret key or incoming e-mail messages. The company openly marketed the service as a way to circumvent FBI surveillance.

But, like e-gold, Hushmail was another formerly crime-friendly service now being mined by law enforcement. U.S. and Canadian agencies had been winning special orders from the Supreme Court of British Columbia that forced Hushmail officials to sabotage their own system and compromise specific surveillance targets' decryption keys. Now the feds had Max's e-mail.

At the same time, the agency had located Tea living in Berkeley serving a probation sentence—it turned out she'd been caught using Aragon-produced gift cards at the Emeryville Apple Store months earlier. It was supposed to be a training run for one of Chris's new recruits, but Tea had never cashed before, and when she impulsively added a PowerBook to her iPod purchase, she was arrested along with the trainee. Eager to avoid more trouble, she'd told the Secret Service everything she knew.

Meanwhile, the Secret Service had begun sporadic physical surveillance of Max. From Werner Janer's proffers, Mularski had learned that Max had a girlfriend named Charity Majors. Public records provided her address, and a subpoena of her bank records showed she had a joint account with Max. The Secret Service staked out the house and eventually trailed Max to the Oakwood Geary.

Electronic surveillance confirmed that Max was operating from the Oakwood. The FBI had won a secret court order letting them electronically monitor the IP addresses connecting to Carders Market's false front at a U.S. hosting company—the modern equivalent of taking down the license plates outside a mob hangout. Several traced back to broadband subscribers living within a block of the corporate apartment complex and running Wi-Fi.

Two weeks earlier, a female Secret Service agent disguised as a maid had ridden up the elevator with Max and watched him unlock apartment 409. The apartment number was the last piece of data they'd needed.

There was just one more stop before they'd move in: the Orange County Central Men's Jail, a grim lockup in the flat, sun-baked center of Santa Ana, California. McKenzie and federal prosecutor Luke Dembosky were shown to an interview room to meet Chris Aragon.

Chris was the last holdout in the Orange County crew. Clara and six members of his crew were headed to plea deals that would ultimately net them from six months to seven years in prison. Clara would get two years and eight months. Chris's mother was looking after the two boys.

Once the introductions were made, McKenzie and Dembosky got down to business. They couldn't do anything about Chris's state case, but if he cooperated, he'd have a nice letter in his file from the U.S. government attesting that he'd helped in a major federal prosecution. That could sway the judge at sentencing time. It was all they could do.

McKenzie produced a photo lineup and asked Chris if anyone looked familiar.

Chris's situation was grim. With his bank robberies and drug-smuggling convictions, he was eligible for California's tough three-strikes law. That meant a mandatory twenty-five-to-life.

Chris picked out Max's mugshot from the photos. And then he told the feds the story of Max Vision's drift to the dark side.

. . .

On Wednesday, September 5, 2007, Max dropped Charity at the post office on an errand and directed his cab driver downtown to the CompUSA store on Market Street. He picked up a new fan for his CPU, walked to his apartment, stripped down, and crashed out on his bed amid a tangle of unfolded laundry. He settled into a deep slumber.

Max had stopped hacking, but he was still disentwining himself from his double life—after five years, he had a lot of relationships and ventures that he couldn't just sever overnight.

He slept right through the knock at his door at about two p.m. Then the door flew open, and a half-dozen agents rushed into the room, guns drawn, shouting orders. Max bolted upright and screamed.

"Put your hands where I can see them!" an agent yelled. "Lay down!" The agent was positioned between Max and his computers. Max had often thought that, in a raid, he might be able to pull the plug on his server, making his already formidable cyberdefenses completely bulletproof. Now that it was really happening, he realized that diving for his machines wasn't an option, unless he wanted to be shot.

Max recovered his composure. Unplugged or not, his machines were locked down, and his encryption was rock solid. He managed to relax a little as the agents let him get dressed, then walked him down the hall in handcuffs.

On the way, they passed a three-man team who'd been waiting for the Secret Service to secure the safe house. They weren't feds; they were from Carnegie Mellon University's Computer Emergency Response Team, and they were there to bust Max's crypto.

It was the first time CERT had been invited to a raid—but the circumstances were special. Chris Aragon had employed the same DriveCrypt whole-disk-encryption software that Max used, and neither the Secret Service nor CERT had been able to recover anything from the drive.

Full-disk encryption keeps the entire hard drive encrypted at all times: all the files, the file names, the operating system, the software, the directory structure—any clue to what the user has been doing. Without the decryption key, the disk might as well have been a Frisbee.

The key to cracking a full-disk encryption program is to get at it while it's still running on the computer. At that point, the disk is still fully encrypted, but the decryption key is stored in RAM, to allow the software to decrypt and encrypt the data from the hard drive on the fly.

The knock on Max's door had been intended to draw Max away from his machines; if he'd shut them down before the Secret Service got the cuffs on, there wouldn't have been much CERT could do—the contents of the RAM would have evaporated. But Max had been caught napping, and his servers were still running.

CERT had spent the last two weeks gaming out different scenarios for what they might encounter in Max's safe house. Now the team leader looked over the setup: Max's server was wired to half a dozen hard drives. Two had lost power when an agent tripped over an electrical cable snaking across the floor, but the server itself was still running, and that was what mattered.

While Secret Service flashbulbs bounced off the walls of Max's cluttered apartment, the forensics experts moved to the machines and began their work, using memory-acquisition software they'd brought with them to suck down the live data from the RAM onto an external storage device.

Down the hall, Max cooled his heels in the feds' apartment.

Two agents watched over him. Max would be questioned later—for now, the agents were just babysitting, chatting with one another. The Secret Service agent was from the local San Francisco field office; he asked his FBI counterpart where he worked.

"I'm from Pittsburgh," Keith Mularski answered.

Max's head snapped to look at Master Splyntr. There was no doubt who had won the carder war.

The Secret Service agents exulted over the bust. "I've been dreaming about you," agent Melissa McKenzie said as she drove Max to the field office. On seeing his raised eyebrow, she added, "I mean about Iceman. Not you personally."

Two of the local agents were dispatched to Charity's house. They told her what happened and took her downtown to say good-bye to Max.

"I'm sorry," he told her when she walked in. "You were right."

Max talked to the agents at the field office for a while, trying to feel them out for what they knew and gauge how much trouble he was in. Some of them seemed surprised at his politeness—his sheer likability. Max wasn't what they expected from the cold, calculating kingpin they'd been tracking for a year.

On the drive to jail, McKenzie finally voiced her puzzlement. You seem like a nice guy, she said, and that's going to help you. "But I have this one question for you. . . .

"Why do you hate us?"

Max was speechless. He never hated the Secret Service, or the FBI, or even the informants on Carders Market. Iceman did. But Iceman was never real; he was a guise, a personality Max slipped on like a suit when he was in cyberspace.

Max Vision never hated anyone in his life.

The Hungry Programmers were the first to hear the news that Max had been arrested again. Tim Spencer offered to sign for Max's bail bond. For collateral, he had twenty acres of land in Idaho that he'd bought as his dream retirement property. When Tim heard the details of the charges against his old friend, he hesitated. What if he didn't really know Max at all?

The moment of doubt passed, and he signed the form. Max's mother offered to post the equity in her house as well to secure her son's release. Ultimately, though, it didn't matter. When Max came up for arraignment in San Jose, a federal magistrate ordered the hacker held without bail pending his transport to Pittsburgh.

The government announced Iceman's arrest on September 11, 2007. The news hit Carders Market, sparking a flurry of activity. Achilous immediately deleted the entire database of posts and private messages, not knowing the feds already had it. "I think the SQL database almost had a heart attack when I did it, but it's done now. I think this is what Aphex would have wanted," he wrote. "This forum is open for posting, so people can chat and figure out where to go from here. Just be very careful, specifically about following links. Try to keep the conspiracy theories to a minimum everyone, please.

"Good luck, be safe."

Silo jumped in under an alias to wrongly label his former rival a snitch, based on news reports that misunderstood Max's work for the FBI during his white-hat days. "It's sad to see a brilliant guy go," he wrote. "He brought a lot to this board and the scene as a vendor and an administrator. A lot of guys made a lot of money from him."

But "once a rat always a rat," he wrote, with no trace of irony. "This whole board is spawned out of the fact that years ago the FBI and Aphex had a disagreement on whom he was snitching out. . . . Bottom line, he is the biggest hypocrite to ever grace the scene."

Back at his desk in Pittsburgh, Mularski put on Master Splyntr's black hat to join the postgame analysis. The FBI agent knew full well that Iceman hadn't been an informant, but his alter ego would be expected to seize on the news that Max had once worked with the feds. "Oh just where do I even begin?" He gloated on DarkMarket, enjoying the moment. "Let's see . . . let's see . . . How about with this headline from SFGate .com? And I quote, 'Ex-FBI snitch in S.F. indicted in hacking of financial institutions.'

"Did anyone else notice anything about that headline? Ahh yeah, FBI Snitch. This is turning out to be just like Gollumfun and El. No wonder why Iceman always had a hard-on for them, because he was just like them and was competing for his handlers' praises."

When Max arrived in Pittsburgh, his new public defender tried again to get him released on bail, but the judge refused after prosecutors speculated that Max was sitting on vast stores of hidden cash and could easily use his contacts to disappear with a new name. To prove that he'd tried to evade the feds, they played their trump card: private messages written by Max himself describing his use of false IDs while traveling and his "evasive move" to his final safe house. Max had sent the messages to a Pittsburgh Secret Service informant who'd been an admin on Carders Market for a full year.

Max wasn't at all surprised to see that it was Th3C0rrupted0ne.

34

DarkMarket

The man is sitting rigid on a polished wooden chair and staring balefully into the camera. Paint peels from a cracked plaster wall behind him. He's been stripped down to his underwear, and he's holding a handwritten sign over his exposed paunch. I AM KIER, it reads, in large block letters. MY REAL NAME IS MERT ORTAC. . . . I AM RAT. I AM PIG. I AM FUCKED BY CHA0.

The appearance of the photo on DarkMarket in May 2008 sent Mularski hurrying back into the NCFTA communications room. Headquarters would want to know that one of Master Splyntr's admins had just kidnapped and tortured an informant.

Cha0 was an engineer in Istanbul who sold high-quality ATM skimmers and PIN pads to fraudsters around the world. Covertly affixed to a cash machine, the skimmer would record the magstripe data on every debit or credit card fed into the ATM, while the PIN-pad overlay stored the user's secret code.

Cha0 cut a jaunty presence in the underground. His Flash-animated banner ad on DarkMarket was a classic, opening with a cartoon man wading through a house full of cash. "Is that you?" the text asks. "Yes. If you bought a skimmer and PIN pad from Cha0." A similarly styled video tutorial for new customers was narrated by a smiling caricature of Cha0 himself. "Hi, my name is Cha0. I'm a developer of skimming devices. I work for you twenty-four hours a day and make the best devices for skimming. You'll be able to make money in this business with me and my group. We make these devices for newbies—it's that easy to use!" The animated Cha0 goes on to offer practical advice: Don't install your skimmer in the morning, because passersby are more vigilant at that time. Don't choose a location where 250 people or more pass a day. Avoid cities with a population less than 15,000—residents know too well what the ATM is supposed to look like and might notice Cha0's product.

Notwithstanding his whimsical marketing, Cha0 had always made it clear to his friend Master Splyntr that he was a serious criminal, not afraid to get physical to protect his multimillion-dollar business. Now he'd proven it. Mert "Kier" Ortac had been part of Cha0's organization, the Crime Enforcers, until he went running to a Turkish TV station to blab about Cha0's activities. After a couple of interviews, he vanished. When he resurfaced a short time later, he told a harrowing story about being abducted and beaten by Cha0 and his henchmen.

Now Cha0 had confirmed the tale by posting the kidnap photo to DarkMarket as a warning to others.

The image put proof to the FBI's long-held suspicions that the computer underground was getting violent. With hundreds of millions of dollars pouring into the scene every year, it had seemed inevitable that the carders would take on the brutal methods of traditional organized crime to enlarge or protect their illegal income.

With Max safely locked up in an Ohio detention center, DarkMarket had been free to grow, and Mularski was closing in on its heaviest hitters—Cha0 among them. A Turkish cybercrime detective had spent three months at the NCFTA on a fellowship and was working with Mularski to run down the skimmer maker.

Mularski had sent Cha0 two lightweight PCs as a gift the previous year, opening the first door in the investigation. Cha0 had directed the shipment to flunkies in his organization, who were promptly put under surveillance by the Turkish National Police. That led to Cagatay Evyapan, an electrical engineer with a prior criminal record—details that jibed with the biography Cha0 had shared privately with Mularski.

The police approached several international shipping companies and briefed them about Cha0's operations. One of them identified some of the skimmer shipments from Istanbul to Europe, fingering a known member of Cha0's organization as the shipper.

That gave the police the evidence they needed. On September 5, five police in bulletproof vests raided Cha0's apartment on the outskirts of Istanbul. They rushed into his house and pushed Cha0 and an associate to the ground at gunpoint.

Inside his apartment was a complete electrical lab and assembly line, with components neatly organized in trays and bins. Nearly a dozen computers were running on the desks. Cha0 had all the same card-counterfeiting equipment that had graced Chris Aragon's factory, as well as giant cardboard boxes holding some one thousand skimmers and two thousand PIN pads, all awaiting international shipment. Cha0's records showed that four of them had already gotten into the United States.

The cops brought Evyapan out in handcuffs, a tall, beefy man with close-cropped hair and a black T-shirt emblazoned with the Grim Reaper. The face of organized crime in the Internet age.

Cha0 was the last listed target in Mularski's undercover authorization; the other key DarkMarket players had already been taken down. Markus Kellerer, Matrix001, was arrested in Germany in May 2007 and spent four months in a high-security prison. Renukanth "JiLsi" Subramaniam, a Sri Lankan–born British citizen, was raided in London in June 2007 after detectives with the Serious Organised Crime Agency in Britain staked out the Internet café he used as an office, matching his appearances at the Java Bean with JiLsi's posts on DarkMarket and his chats with Master Splyntr. JiLsi's associate, sixty-seven-year-old John "Devilman" McHugh, was picked up at the same time; police found a credit card counterfeiting factory in the senior citizen's home.

In Turkey, six members of Cha0's organization were charged along with Cha0. With Mularski's help, the police also swooped in on Erkan "Seagate" Findikoglu, a DarkMarket member who ran a massive King Arthur–style cash-out operation responsible for at least two million dollars in thefts from U.S. banks and credit unions—they recovered one million of it in cash at his arrest. Twenty-seven members of Seagate's organization were charged in Turkey, and the FBI rounded up six of his cashers in the United States.

With Cha0 and Seagate in jail, Mularski's work was done—his two years running DarkMarket had now resulted in fifty-six arrests in four countries. On Tuesday, September 16, 2008, he drafted a post formally announcing the closure of the site. As an homage to the carding world's history and culture, the FBI agent borrowed from King Arthur's legendary message closing Carder Planet years before. "Good day, respected and dear forum members," he began.

> It is time to tell you the bad news—the forum should be closed. Yes, I really mean closed.
>
> Over the last year we have lost a lot of the admins of the forums: Iceman on Carders Market; JiLsi and Matrix001 disappeared, and now, Cha0 on DM. It is apparent that this forum, which has been around almost three years, is attracting too much attention from a lot of the world services. . . .
>
> I myself would rather go out like King Arthur than Iceman. Whereas Iceman decided that all he would do was change his nick to Aphex,

> and continue to run CM, King Arthur closed CarderPlanet and faded into the night. History has shown that Iceman made a fatal mistake. I will not make the same.

Mularski planned to keep his Master Splyntr identity dormant but alive: He'd have a well-established underground legend that he could pull from his pocket whenever he needed it in future investigations. But it was not to be. About a week after DarkMarket went dark, a reporter for Südwestrundfunk, Southwest Germany public radio, got his hands on court documents filed in Matrix's case that laid bare Mularski's double life. The U.S. press picked up the story. Now 2,500 members of DarkMarket knew they'd been doing business on a sting site and that Iceman had been right all along.

Three days after the story broke in the United States, Mularski found an ICQ message to Master Splyntr waiting on his computer. It was from TheUnknown, a UK target who'd gone on the run after he was raided by the British police. "U fucking piece of shit. Motherfucker. Thought you can catch me. Hahaha. Fucking newb. U are nowhere near me."

"If you want to make arrangements to turn yourself in, let me know," Mularski wrote back. "It will be easier than looking over your shoulder the rest of your life."

TheUnknown turned himself in a week later.

Mularski was almost relieved to have his secret identity revealed; for two years, his laptop had been his constant companion—even on vacation, he'd been online talking to carders. He'd enjoyed some of it—building online friendships with some of his targets, teasing and taunting others. Master Splyntr could say things to criminals that a respectable FBI agent never could.

Eager as Mularski was to have his life back, it would take time. Nearly a month after DarkMarket's closing, he was still fighting a vague restlessness. Mularski had one more challenge to master. He'd have to learn how to not be Master Splyntr.

35

Sentencing

Max towered over the marshals as they brought him into the Pittsburgh courtroom to face sentencing. He wore an ill-fitting orange jail uniform, his hair trimmed short and neat.

His escorts uncuffed his hands, and he took a seat next to his public defender at the defense table. A half-dozen reporters talked among themselves on one side of the gallery, an equal number of feds on the other. Behind them, the long wooden pews were mostly empty: no friends, no family, no Charity; she'd already told Max she wasn't going to wait for him.

It was February 12, 2010, two and a half years after his arrest at the safe house. Max had spent the first month locked up at the Santa Clara county jail, speaking daily with Charity in long phone calls more intimate than any conversations they'd had while he was immersed in his crimes. The marshals finally put him on a plane and checked him into a detention facility in Ohio, where Max made peace with his confinement, largely drained now of the self-righteous anger that carried him through his previous imprisonments. He made new friends in the joint: geeks like him. They started a Dungeons and Dragons campaign.

By year's end, Max had no more secrets. It had taken the CERT investigators only two weeks to find the encryption key in the image of his computer's RAM. At one of his court appearances, prosecutor Luke Dembosky handed Max's lawyer a slip of paper with his passphrase written on it: "!!One man can make a difference!"

For years, Max had used his encrypted hard drive as an extension of his brain, storing everything he found and everything he did. That the feds had it was disastrous for his legal future, but more than that, it felt like an intimate violation. The government was in his head, reading his mind and memories. When he returned to his cell after the hearing, he wept into his pillow.

They had everything: five terabytes of hacking tools, phishing e-mails, dossiers he'd compiled on his online friends and enemies, notes on his interests and activities, and 1.8 million credit cards accounts from over a thousand banks. The government broke it down: Max had stolen 1.1 million of the cards from point-of-sale systems. The remainder mostly came from the carders Max had hacked.

It was eight miles of magstripe data, and the feds were prepared to charge him for every inch. The government had secretly flown Chris to Pittsburgh for weeks of debriefing while the credit card companies tallied the fraudulent charges on Max's cards, arriving at a staggering $86.4 million in losses.

Max's profits were far less: Max told the government he earned under $1 million from his capers and had pissed most of it away on rent, meals, cab fare, and gadgets. The government found about $80,000 in Max's WebMoney account. But federal sentencing guidelines in theft cases are based on victim harm, not the offender's profits, so Max could be held responsible for the charges rung up by Chris, the carders who bought dumps from Digits and Generous, and potentially the fraud performed by the carders Max hacked. Rolled up with Max's rap sheet, the $86 million translated to a sentence of thirty years to life, with no parole.

Faced with decades in prison, Max began cooperating with the investigation. Mularski took him out for long debriefing sessions about the hacker's crimes. At one of them, after the DarkMarket sting broke in the press, Max apologized to Mularski for his attempts to expose Master Splyntr. Mularski heard sincerity in his old foe's voice and accepted his apology.

After a year of negotiation, Max's lawyer and the government settled on their number—a joint recommendation to the judge of thirteen years. In July 2009, Max had pleaded guilty.

The deal wasn't binding on the court; in theory, Max could be released on the spot, sentenced to life, or anything in between. The day before the sentencing, Max typed out a four-page letter to his judge, Maurice Cohill Jr., a seventy-year-old Ford appointee who'd been a jurist since before Max was born.

"I don't believe further prison time in my case will help anyone," Max wrote. "I don't think it is necessary because all I want to do is help. I disagree with the blanket assessment of the sentencing guidelines. Unfortunately, I am facing such a horrible sentence that even 13 years seems 'good' in comparison. But I assure you it is overkill as I am the proverbial dead horse. That said, I plan to make the most of the time I have left on this earth be it in prison or otherwise."

He continued. "I have a lot of regrets, but I think my essential failing was that I lost touch with the accountability and responsibility that comes with being a member of society. A friend of mine once told me to behave as though everyone could see what I was doing all the time. A sure way to avoid engaging in illegal conduct, but I guess I wasn't a believer because when I was invisible, I forgot all about this advice. I know now that we can't be invisible, and that it's dangerous thinking."

Max watched with studied calmness as his lawyer stood to confer with the prosecution over last-minute details and the courthouse staff went through their prehearing checklist, testing the microphones and shuffling papers. At ten thirty a.m. the door to chambers opened. "All rise!"

Judge Cohill took the bench. A wizened man with a close-cropped snow-white beard, he peered at the courtroom through round glasses and announced the sentencing of Max Butler, the name under which Max had been charged. He read Max's sentencing guidelines for the record, thirty years to life, then listened as prosecutor Dembosky laid out his case for leniency. Max had provided significant help to the government, he said, and was deserving of a sentence below the guidelines.

What followed could have been an awards presentation instead of a sentencing hearing, with Max's lawyer, prosecutor, and judge taking turns praising Max's computer skills and apparent remorse. "He's an extremely bright, self-taught computer expert," said federal public defender Michael Novara, albeit one who orchestrated "computer security breaches on a grand scale."

Dembosky, a computer-crime specialist and seven-year veteran of the U.S. Attorney's Office, called Max "extremely bright and articulate and talented." He'd been at some of Max's debriefings, and like virtually everyone who knew Max in real life, he'd grown to like the hacker. "He's almost wide-eyed and optimistic in his view of the world," he said. Max's cooperation, he added, was why they were asking for only thirteen years instead of an "astronomical" sentence. "I believe that he is very sorry."

Max had little to add. "I've changed," he said. Hacking no longer held any appeal for him. He invited Judge Cohill to ask him any questions. Cohill didn't need to. The judge said he was impressed by Max's letter and by letters written by Charity, Tim Spencer, and Max's mother, father, and sister. He was satisfied that Max was remorseful. "I don't think I have to give you a lecture on the problems you've caused for your victims."

Cohill had already written the sentencing order. He read from it aloud. Thirteen years in prison. Max would also be responsible for $27.5 million in restitution, based on the cost to the banks of reissuing the 1.1 million cards Max stole from point-of-sale systems. Upon his release, he'd serve five years of court supervision, during which he'd be allowed to use the Internet only for employment or education.

"Good luck," he said to Max.

Max stood up—his face neutral—and let a marshal handcuff him behind his back, then lead him through the door in the back of the courtroom connecting to the holding cells. With credit for time served and good behavior, he'd be out just before Christmas 2018.

Almost nine years in prison were still ahead of him. At the time it was the longest U.S. sentence ever handed out to a hacker.

36

Aftermath

By the time Max was sentenced, the Secret Service had identified the mystery American hacker who'd made Maksik into the world's top carder, and he was poised to get a sentence that would make Max's look like a traffic fine.

The big break in the case came from Turkey. In July 2007, the Turkish National Police learned from the Secret Service that Maksik, twenty-five-year-old Maksym Yastremski, was vacationing in their country. An undercover Secret Service operative lured him to a nightclub in Kemer, where police arrested Yastremski and seized his laptop.

The police found the laptop hard drive impenetrably encrypted, just as when the Secret Service performed its sneak-and-peek in Dubai a year earlier. But after a few days in a Turkish jail, Maksik coughed up the seventeen-character passphrase. The police gave the passphrase and a copy of the disk to the Secret Service, which began poring over its contents, taking particular interest in the logs Maksik kept of his ICQ chats.

One chat partner stood out: ICQ user 201679996 could be seen helping the Ukrainian with a hack attack against the restaurant chain Dave & Buster's and discussing some of the earlier high-profile intrusions that had put Maksik on the map. The agents checked out the ICQ number and obtained the e-mail address used to first register the account: soupnazi @efnet.ru.

SoupNazi was a name the agency had heard before—in 2003, when they arrested Albert Gonzalez.

Gonzalez was the informant who'd lured Shadowcrew carders into a wiretapped VPN, leading to the twenty-one arrests in Operation Firewall—the Secret Service's legendary crackdown on the carding scene. But years before he was known as Cumbajohnny on Shadowcrew, Gonzalez had used the *Seinfeld*-inspired handle SoupNazi in IRC.

The carder turncoat who'd made Operation Firewall possible had gone on to stage the largest identity thefts in U.S. history.

One month after Firewall, Gonzalez had gotten permission to move from New Jersey back to his home, Miami, where he'd launched the second act of his hacking career. He took on the name Segvec and passed himself off as a Ukrainian, hanging his hat on the Eastern European forum Mazafaka. Under the rubric Operation Get Rich or Die Tryin'—the title of a 50 Cent album and Maksik's Shadowcrew motto—he went on to create a multimillion-dollar cybertheft ring that touched tens of millions of Americans.

On May 8, 2008, the feds swooped in on Gonzalez and his U.S. associates. Hoping for leniency at sentencing, Gonzalez cooperated again, providing agents with the encryption key for his hard drive and giving them information on his entire gang. He admitted to the breaches at TJX, OfficeMax, DSW, Forever 21, and Dave & Buster's, and to helping Eastern European hackers penetrate the grocery chain Hannaford Bros., 7-Eleven's ATM network, Boston Market, and the credit card processing company Heartland Payment Systems, which alone leaked nearly 130 million cards. It was a lucrative business for the hacker. Gonzalez drew the Secret Service a map to over $1 million in cash he'd buried in his parents' backyard; the government sought forfeiture of the money, his 2006 BMW, and a Glock 27 firearm with ammunition.

Gonzalez had built his crew from an untapped reservoir of hacker talent—onetime bedroom hackers who had trouble finding a place in the white-hat world. Among them was Jonathan "C0mrade" James, who'd hacked NASA as a teenager and received a landmark six-month juvenile sentence the same week Max Vision pleaded guilty to his Pentagon hacks in 2000. After a brief flurry of fame—including an interview on PBS's *Frontline*—James slipped into obscurity, living quietly in a house he inherited from his mother in Miami.

Then in 2004 he allegedly began working with Gonzalez and an associate named Christopher Scott. The government believes James and Scott were responsible for one of the earliest magstripe hauls to make their way into Maksik's vaults, cracking OfficeMax's Wi-Fi from a store parking lot in Miami and stealing thousands of swipes and encrypted PINs. The two allegedly provided the data to Gonzalez, who arranged with another hacker to decrypt the PIN codes. Credit card companies later reissued some two hundred thousand cards in response to the attack.

Of all the hackers, it was Jonathan James who would pay the highest price in the post-Shadowcrew carder crackdown. In the days after his May 2008 raid, James became convinced the Secret Service would try to pin all of Gonzalez's breaches on him to wring public relations juice out of his notorious past and protect their informant, Gonzalez. On May 18, the twenty-four-year-old stepped into the shower with a handgun and shot himself dead.

"I have no faith in the 'justice' system," read his five-page suicide note. "Perhaps my actions today, and this letter, will send a stronger message to the public. Either way, I have lost control over this situation, and this is my only way to regain control."

In March 2010, Gonzalez was sentenced to twenty years in prison. His U.S. coconspirators drew sentences ranging from two to seven years. In Turkey, Maksik was convicted of hacking Turkish banks and sentenced to thirty years.

Since Max's arrest, new scams have emerged in the underground, the worst of them involving specialized Trojan horse software designed to steal a target's online banking passwords and initiate money transfers from the victim's account right through his own computer. The thieves have devised an ingenious solution to the problem that had bedeviled Chris Aragon: how to get at the money. They recruit ordinary consumers as unwitting money launderers, dangling bogus work-at-home opportunities, in which the "work" consists of accepting money transfers and payroll deposits, then sending the bulk of the cash to Eastern Europe by Western Union. In 2009, the scheme's first year of widespread operation, banks and their customers lost an estimated $120 million to the attack, with small businesses the most common target.

Meanwhile, the sale of dumps continues, dominated now by a new crop of vendors, same as the old crop—Mr. BIN; Prada; Vitrium; The Thief.

Law enforcement, though, has claimed some lasting victories. So far, no prominent English-speaking board has risen to replace Carders Market and DarkMarket, and the Eastern Europeans have become more cloistered and protective. The big players have retreated to invitation-only encrypted chat servers. The marketplace exists, but the carders' sense of invulnerability is shattered, and their commerce is tariffed by paranoia and mistrust, thanks primarily to the FBI, the Secret Service, their international partners, and the unheralded work of the post office.

The veil of secrecy that once protected hackers and corporations alike has mostly evaporated, with law enforcement no longer going out of its way to shield companies from responsibility for their poor security. More than one of Gonzalez's hacking targets were made public for the first time in his federal indictment.

Finally, Mularski's DarkMarket sting proved the feds don't have to get in bed with the bad guys to make busts.

All the lowest moments in the war on the computer underground came about through the antics of informants. Brett "Gollumfun" Johnson, the snitch who briefly worked as a Carders Market administrator, turned the Secret Service's Operation Anglerphish into a circus by staging a tax refund scam on the side. Albert Gonzalez provided the clearest example. After Operation Firewall, the Secret Service had been paying Gonzalez an annual salary of $75,000 a year, even as he staged some of the largest credit card hacks in history.

The post-Shadowcrew magstripe breaches led to a reckoning in the civil courts. TJX paid $10 million to settle a lawsuit filed by the attorneys general of 41 states and another $40 million to Visa-issuing banks whose cards were compromised. Banks and credit unions filed lawsuits against Heartland Payment Systems for the massive breach at the transaction-processing firm. Gonzalez's attacks also tore a hole in the credit card industry's primary bulwark against breaches: the so-called Payment Card Industry—or PCI—Data Security Standard, which dictates the steps merchants and processors must take to protect systems handling credit card data. Heartland had been certified PCI compliant before it was breached, and Hannaford Brothers won the security certification even as hackers were in its systems, stealing credit card swipes.

When the dust began to settle from Gonzalez's large-scale hacks, the smaller but far more numerous attacks against restaurant point-of-sale systems began to come out. Seven restaurants in Mississippi and Louisiana who'd suffered intrusions figured out they were all using the same point-of-sale system, the Aloha POS that was once Max's favorite target. The restaurants filed a class-action lawsuit against the manufacturer and the company that sold them the terminals, Louisiana-based Computer World, which allegedly installed the remote-access software pcAnywhere on all the machines and set the passwords on all of them to "computer."

Underlying all these breaches is a single systemic security flaw, exactly 3.375 inches long. Credit card magstripes are a technological anachronism, a throwback to the age of the eight-track tape, and today the United States is virtually alone in nurturing this security hole. More than a hundred other countries around the globe, in Europe, Asia, and even Canada and Mexico, have implemented or begun phasing in a far more secure system called EMV or "chip-and-PIN."

Instead of relying on a magstripe's passive storage, chip-and-PIN cards have a microchip embedded in the plastic that uses a cryptographic handshake to authenticate itself to the point-of-sale terminal and then to the transaction-processing server. The system leaves nothing for a hacker to steal—an intruder sitting on the wire could eavesdrop on the entire transaction and still be unable to clone a card, because the handshake sequence changes every time.

White hats have devised attacks against chip-and-PIN, but nothing that would lend itself to the mass market in dumps that still exists today. So far, the biggest flaw in the system is that it supports magstripe transactions as a fallback for Americans traveling abroad or tourists visiting the United States.

American banks and credit card companies have rejected chip-and-PIN because of the enormous cost of replacing hundreds of thousands of point-of-sale terminals with new gear. In the end, the financial institutions have decided their fraud losses are acceptable, even with the likes of Iceman prowling their networks.

EPILOGUE

In the Orange County men's jail, Chris Aragon is lonely, feeling abandoned by his friends and torn with grief that his children are growing up without him. In October 2009, Clara filed for divorce, seeking custody of their two children. His girlfriend filed for child support.

Chris is studying the *Bhagavad Gita* and has a full-time job as an inmate representative, helping several hundred prisoners with legal matters, medical complaints, and issues with the jail staff. His lawyer is playing a waiting game, winning endless continuances for the criminal trial that, if he loses, still carries a twenty-five-to-life term. After Chris's story was featured in a *Wired* magazine article on Max, Chris was contacted by a Hollywood screenwriter and a producer, but he didn't respond. His mother suggested he get an agent.

Max was assigned to FCI Lompoc, a low-security prison an hour north of Santa Barbara, California. He hopes to use his time to get a degree in physics or math—finally completing the college education that was interrupted a decade earlier in Boise.

He's taken a mental inventory and is dismayed to find that, despite everything, he still has the same impulses that guided him into a life of hacking. "I'm not sure how to really mitigate that, except ignore it," he said in an interview from jail. "I really believe that I'm reformed. But I don't know what's going to happen later."

It might seem a curious confession—admitting that the elements of his personality that landed him in prison still remain buried deep inside. But Max's new self-awareness shows hope for real change. If one is born a hacker, no amount of prison can drive it out. No therapy, or court supervision, or prison workshop can offer reform. Max has to reform himself—learn to own his actions and channel the useful parts of his nature into something productive.

To that end, Max has volunteered to help the government during his confinement, defending U.S. networks or perhaps counterattacking foreign adversaries online. He wrote out a menu of the services he could offer in a memo headed "Why the USA Needs Max." "I could penetrate China's military networks and military contractors," he suggested. "I can hack al Qaida." He's hopeful he might do enough for the government that he could apply for a lowered sentence from his judge.

It's a long shot, and so far, the feds haven't taken him up on his offer. But a month after his sentencing, Max took a baby step in that direction. Keith Mularski arranged for Max to speak at the NCFTA for an eager audience of law enforcement officials, students, financial and corporate security experts, and academics from Carnegie Mellon.

Mularski checked him out of jail for the appearance. And for an hour or two, Max Vision was a white hat again.

NOTES

Prologue

xi *The taxi idled:* Interviews with Max Vision.

Chapter 1: The Key

1 *As soon as the pickup truck rolled up to the curb:* Interviews with Max's friend Tim Spencer. The confrontation was also described in less detail by Kimi Mack, Max's ex-wife. Though Max could intimidate bullies, he was never forced into a physical confrontation with them.

2 *Max's parents had married young: State of Idaho v. Max Butler,* 1991. District Court of the Fourth Judicial District, Ada County, Case No. 17519.

2 *Robert Butler was a Vietnam veteran: State of Idaho v. Max Butler* and interviews with Max.

2 *Weather Channel and nature documentaries:* Interviews with Kimi Winters and Max, respectively. Max's parents declined to be interviewed.

2 *relaxed, and full-bore insane:* Interviews with Tim Spencer and with "Amy," Max's ex-girlfriend. Max's emotional problems at this time are also reflected in court records in *State of Idaho v. Max Butler.* Max acknowledges that his parents' divorce had a deep effect on him.

2 *One day he emerged from his home:* Interview with Tim Spencer. Max confirms the incident but says he lit the fire in a field adjacent to Spencer's house.

2 *The Meridian geeks had found the key ring:* The account of the master-key incident comes from interviews with Tim Spencer. Court records confirm Max's juvenile conviction. Max admits the trespass and chemical theft but declined to detail

what occurred inside the school. John, his uncharged accomplice in the burglary, declined comment.

4 *Max became "Lord Max":* Max described his run-in with the Secret Service in an interview. Also referenced in a letter Max wrote that was filed in *State of Idaho v. Max Butler.*

Chapter 2: Deadly Weapons

6 *THIS is the Rec Room!!!!: From MUDs to Virtual Worlds,* Don Mitchell, Microsoft social computing group (March 23, 1995).

6 *three hundred thousand host computers:* Numerous sources, including *"Illuminating the net's Dark Ages,"* Colin Barras, BBC News, August 23, 2007.

7 *At Max's urging:* The events surrounding Max's assault conviction are based on transcripts and other documents in *State of Idaho v. Max Butler,* as well as interviews with Max and "Amy." Where there are significant factual disputes, they are noted herein.

8 *Then the dark truth:* "The Dreaming City," Michael Moorcock, *Science Fantasy* 47 (June 1961).

9 *Like a few of them, he started hacking the computer right away:* The hacking at BSU was described by Max and David in interviews. David described Max's speed and impatience. BSU professor Alexander Feldman discussed Max's computer ban in an interview and said Max had probed other computers.

10 *The sheriff called BSU's network administrator at two in the morning:* Interview with Greg Jahn, a former BSU system administrator responsible for locking down Max's account and preserving his files.

Chapter 3: The Hungry Programmers

15 *Idaho's Supreme Court ruled: State v. Townsend,* 124 Idaho 881, 865 P.2d 972 (1993).

16 *Max found an unprotected FTP file server: Cinco Network, Inc. v. Max Butler,* 2:96-cv-1146, U.S. District Court, Western District of Washington. Max confirms this account but says he was primarily interested in distributing music files, not pirated software.

18 *Chris Beeson, a young agent:* The details of Max's assistance to the FBI come from court filings by the defense attorney in his subsequent criminal case, *USA v. Max Ray Butler,* 5:00-cr-20096, U.S. District Court, Northern District of California. Details of his recruitment and his relationship with the agents come from interviews with Max and Max's Internet writings immediately following his

guilty plea. See http://www.securityfocus.com/comments/articles/203/5729/threaded (May 24, 2001). Max says he did not consider himself an informant and only provided technical information.

Chapter 4: The White Hat

19 *The first people to identify themselves as hackers:* The seminal work on the early hackers is Steven Levy, *Hackers: Heroes of the Computer Revolution* (New York: Anchor Press/Doubleday, 1984). Also see Steve Wozniak and Gina Smith, *iWoz: From Computer Geek to Cult Icon: How I Invented the Personal Computer, Co-Founded Apple, and Had Fun Doing It* (New York: W. W. Norton and Company, 2006).

21 *Tim was at work one day:* This anecdote was recalled by Tim Spencer. Max later recalled Spencer's advice in a letter to his sentencing judge in Pittsburgh.

21 *If there was one thing Max:* Details of Max's relationship with Kimi come primarily from interviews with Kimi.

23 *Max went up to the city to visit Matt Harrigan:* Harrigan's business and his work with Max were described primarily by Harrigan, with some details confirmed by Max.

Chapter 5: Cyberwar!

26 *In 1998, security experts discovered the latest flaw in the code:* This account of Max's BIND attack draws primarily from court records, including Max's written confession, interviews with Kimi, and interviews with former air force investigator Eric Smith. E-mail snippets between Max and the FBI are from court records. Technical details come primarily from a contemporaneous analysis of Max's code that can be found at http://www.mail-archive.com/redhat-list@redhat.com/msg01857.html.

27 *issued an alert:* "Inverse Query Buffer Overrun in BIND 4.9 and BIND 8 Releases," CERT Advisory CA-98.05.

30 *He sent Paxson an anonymous note:* The note was provided to the author by Vern Paxson. Max confirmed that he sent it.

Chapter 6: I Miss Crime

34 *Kimi came home from school:* Kimi described this portion of the FBI search and its aftermath.

34 *The FBI agents saw an opportunity in Max's crime:* The details come from court

filings by the defense attorney in *USA v. Max Ray Butler,* 5:00-cr-20096, U.S. District Court, Northern District of California.

36 *Max was in heaven:* Interviews with Max and Kimi.

38 *Carlos Salgado Jr., a thirty-six-year-old computer repairman:* Details of the Salgado caper come from interviews with Salgado, Salgado's intended buyer, the former system administrator of the ISP he hacked, and court records in *USA v. Carlos Felipe Salgado, Jr.,* 3:97-cr-00197, U.S. District Court, Northern District of California. The FBI declined to comment on the case or to identify the victim of the credit card breach.

40 *The next day, Max met Harrigan at a Denny's:* Interviews with Matt Harrigan and Max.

Chapter 7: Max Vision

43 *In late 1998, a former NSA cybersecurity:* Interview with Marty Roesch.

45 *The reason I signed the confession:* Interviews with Kimi. In interviews with the author, Max expressed the sentiment that his attachment to Kimi worsened his legal situation.

47 *"It's his stuff":* Snort IDS mailing list, April 3, 2000. (http://archives.neohapsis.com/archives/snort/2000-04/0021.html).

47 *Patrick "MostHateD" Gregory: "Computer Hacker Sentenced,"* U.S. Department of Justice press release, September 6, 2000 (http://www.justice.gov/criminal/cybercrime/gregorysen.htm).

47 *Jason "Shadow Knight" Diekman: "Orange County Man in Federal Custody for Hacking into Government Computers,"* U.S. Department of Justice press release, September 21, 2000 (http://www.justice.gov/criminal/cybercrime/diekman.htm).

47 *Sixteen-year-old Jonathan James: "Juvenile Computer Hacker Sentenced to Six Months in Detention Facility,"* U.S. Department of Justice press release, September 21, 2000 (http://www.justice.gov/criminal/cybercrime/comrade.htm).

Chapter 8: Welcome to America

49 *The two Russians:* The details of the Invita sting and the background of the Russian defendants come primarily from court records, particularly *USA v. Vassily Gorshkov,* 2:00:mj:00561, U.S. District Court, Western District of Washington, as well as an interview with a former FBI agent who worked on the operation. The description of the Russians' attire and the reference to "the Expert Group"

comes from the excellent *Washington Post* story "A Tempting Offer for Russian Pair" by Ariana Eunjung Cha, May 19, 2003. Quotes from within the Invita office come from a transcript of the surveillance tape, with minor grammatical changes for readability.

Chapter 9: Opportunities

54 *Max wore a blazer and rumpled cargo pants:* The author was present at Max's sentencing hearing: see "As the Worm Turns," SecurityFocus, *Businessweek* online, May 21, 2001 (http://www.businessweek.com/technology/content/jul2001/tc20010726_443.htm). The letters written on Max's behalf are filed in *USA v. Max Ray Butler,* 5:00-cr-20096, U.S. District Court, Northern District of California.

56 *Kimi was talking to him on the phone:* Interview with Kimi.

57 *Max took the news with eerie calm:* Interview with Max.

57 *"I've been talking to some people":* Interview with Kimi.

57 *Jeffrey James Norminton:* Three of Norminton's close associates, Chris Aragon, Werner Janer, and an anonymous source, described Norminton's alcoholism, and Aragon discussed its effect on Norminton's criminal productivity. Federal court records show Norminton's assignment to a drug and alcohol rehabilitation center, and local court records reflect two DUI arrests in 1990 (Orange County Superior Court cases SM90577 and SM99355).

57 *Norminton's latest caper: USA v. Jeffrey James Norminton,* 2:98-cr-01260, U.S. District Court, Central District of California.

58 *Norminton made it clear that he saw real potential in Max:* Interviews with Max, Chris Aragon, Werner Janer, and another source familiar with Max's and Norminton's jailhouse planning.

59 *Max refused to sign:* Kimi and Max agree on this. Max says he refused to sign because Kimi appeared to be wavering in her commitment to divorce him.

59 *I have been showing up at places:* Max's plea to the security community is archived at http://seclists.org/fulldisclosure/2002/Aug/257.

59 *Even the Honeynet Project:* Max says the project shunned him. Founder Lance Spitzner did not answer an inquiry from the author.

61 *A global survey:* Conducted by the Belgian computer security company Scanit by way of a free online vulnerability assessment tool, July 9, 2003.

Chapter 10: Chris Aragon

64 *Max met his future friend and criminal partner Chris Aragon:* Chris Aragon provided this account of his first meeting with Max. Max doesn't remember where they first met.

65 *The first robbery:* The first attempted bank robbery and the final successful one are described in court records for *USA v. Christopher John Aragon and Albert Dwayne See,* 81-cr-133, U.S District Court for the District of Colorado. Additional details, including the Dumpster incident and Aragon's lifestyle at the time, come from the author's interviews with Albert See, Aragon's former crime partner. In interviews, Aragon generally acknowledged his bank robbery conviction and his use of cocaine in this period.

66 *he delved into credit card fraud:* Per Aragon, and confirmed by his former associate Werner Janer and Max.

66 *busted in a nationwide DEA undercover operation:* Kathryn Sosbe, "13 arrested in marijuana bust/Colombian cartel used Springs as distribution point," *Colorado Springs Gazette-Telegraph,* September 13, 1991. The Federal Bureau of Prisons confirmed Aragon's conviction and sentencing on a charge of travel in interstate commerce in aid of a business enterprise involving the distribution of marijuana.

67 *They wound up at the twenty-seven-story Holiday Inn:* The descriptions of Max and Aragon's work together here and throughout this book come primarily from interviews with Max and Aragon, as well as their associates Werner Janer, Jonathan Giannone, Tsengeltsetseg Tsetsendelger, and another source involved in their crimes. Statements provided by Jeffrey Norminton to the FBI, summarized in court documents, also confirm many of the details.

68 *a white-hat hacker had invented a sport called "war driving":* "Evil" Pete Shipley. See the author's "War Driving by the Bay," Securityfocus.com, April 12, 2001 (http://www.securityfocus.com/news/192).

69 *Janer offered to pay Max $5,000 to penetrate the computer of a personal enemy:* According to Aragon, Max, and other sources. Janer says the money was a loan. Charity confirms she received the check on Max's behalf.

70 *Charity had only the broadest notion of what Max was up to:* Interviews with Charity Majors.

71 *On a whim, he cracked Kimi's computer:* Interview with Max.

Chapter 11: Script's Twenty-Dollar Dumps

73 *In the spring of 2001, some 150 Russian-speaking computer criminals:* Greg Crabb, U.S. Postal Inspection Service. Roman Vega, currently in U.S. custody, declined comment, as did the Ukrainian widely suspected to be Script.

73 *The discussion was sparked by:* This history of the carding forums comes from interviews with several veteran carders, court records, interviews with law enforcement officials, and a detailed examination of the archives of Counterfeit Library, CarderPlanet, and Shadowcrew.

77 *the CVV began driving down fraud costs immediately:* Fraud figures come from a presentation by Steven Johnson, director, Visa USA Public Sector Sales, at the ninth annual GSA SmartPay Conference in Philadelphia, August 23, 2007.

78 *Chris decided to try some carding himself:* Aragon described his dealings with Script and his first fraudulent purchases.

Chapter 12: Free Amex!

80 *Max broached his plan obliquely with Charity:* Interview with Charity Majors.

81 *Internet Explorer can process more than just Web pages:* Drew Copley and eEye Digital Security, "Internet Explorer Object Data Remote Execution Vulnerability," August 20, 2003. See CERT Vulnerability Note VU#865940. The author located Max's attack code in a 2003 post to a hacker Web forum, and computer security researcher Marc Maiffret, an executive at eEye, confirmed that it exploited this bug. Max remembers having the vulnerability before it was public but isn't sure how he obtained it. He says eEye and its researchers never leaked bugs in advance.

83 *The disk was packed with FBI reports:* Aragon, Max, and Werner Janer all related the story of Max's intrusion into the FBI agent's computer. Max, Janer, and another source confirmed the agent's name. The agent, E. J. Hilbert, insists he was never hacked and that Max likely penetrated an FBI honeypot filled with fake information.

Chapter 13: Villa Siena

86 *Chris loaded blank PVC cards:* Aragon admits his credit card counterfeiting operation and provided some details in interviews. The author examined Aragon's counterfeiting gear, and dozens of his finished cards, at the Newport Beach Police Department. The blow-by-blow on how the equipment operates comes from interviews with another experienced card counterfeiter who used the same gear.

88 *summoned his girls:* Nancy Diaz Silva and Elizabeth Ann Esquere have pleaded guilty for their roles in Aragon's operation. The other cashers were described variously by Aragon's former associates Werner Janer, Jonathan Giannone, and Tsengeltsetseg Tsetsendelger.

88 *They'd be "sticking it to the man":* The Newport Beach Police Department interviewed one of Aragon's later cashers, Sarah Jean Gunderson, in 2007. According to the police report: "Aragon stated that it was 'The man that we are sticking it to.' Gunderson said she knew it was wrong, however all of her bills were getting paid." Gunderson has pleaded guilty.

Chapter 14: The Raid

92 *Chris Toshok awoke to the sound of his doorbell buzzing:* The details of the raid come primarily from Toshok's blog post "The whole surreal story," *I am Pleased Precariously* on January 15, 2004.

95 *The FBI tried to lure Gembe to America:* Cassell Bryan-Low, "Hacker Hitmen," *Wall Street Journal*, October 6, 2003. Also see the author's "*Valve Tried to Trick* Half-Life 2 *Hacker into Fake Job Interview*," Wired.com, November 12, 2008. (http://www.wired.com/threatlevel/2008/11/valve-tricked-h/).

96 *"Call me back when you're not stoned":* Aragon and Max both agree they fought over money. This quote was recalled by Aragon.

97 *sending them to Mexico to be fitted with clean VINs:* Interviews with Werner Janer and Jonathan Giannone. Court records from Aragon's San Francisco arrest show his car was found to have fake VIN tags, and as part of the case settlement Aragon agreed to forfeit the vehicle. Aragon declined to elaborate on that aspect of his activities in interviews.

Chapter 15: UBuyWeRush

98 *Cesar had come to the underground by a circuitous course:* Interview with Carranza.

100 *Selling equipment wasn't in and of itself illegal:* Carranza pleaded guilty to money laundering in December 2009 for running an e-gold exchange service for carders under the UBuyWeRush brand. *U.S. v. Cesar Carranza,* 1:08-cr-0026 U.S. District Court for the Eastern District of New York. On September 16, 2010, he was sentenced to six years in prison.

101 *The midsized Commerce Bank in Kansas City, Missouri, may have been the first:* Interview with Mark J. Tomasic, former vice president of bank card security with

Commerce Bank. Also see "Hey, banks, earn your stripes and fight ATM fraud scams," *Kansas City Star,* June 1, 2008.

102 *Citibank, the nation's largest consumer bank by holdings, was the most high-profile victim:* The CVV attacks were widely known as the "Citibank cash-outs" in carding circles. One of King Arthur's cashers, Kenneth Flury, was prosecuted in the United States after admitting to stealing $384,000 in Citibank ATM withdrawals in ten days in the spring of 2004: *U.S. v. Kenneth J. Flury,* 1:05-cr-00515, U.S. District Court for the Northern District of Ohio. Citibank declined comment. To discourage competitors, masterminds of the cash-outs often claimed to have secret algorithms at their disposal to generate workable magstripes. Max and other carders confirmed this was a myth, as did FBI agent J. Keith Mularski. Any data would work.

103 *once let it slip to a colleague that King was making $1 million a week:* Joseph Menn, "Fatal System Error," *Public Affairs,* January 2010.

103 *Max had passed them all to Chris, who tore into them with a vengeance:* Interview with Max. Werner Janer confirmed that Chris worked on the Citibank cash-outs with Max, but Janer did not know the details. Aragon declined to comment on the cash-outs.

104 *In just one year:* Avitan Litan, "Criminals Exploit Consumer Bank Account and ATM System Weaknesses," Gartner report G00129989, July 28, 2005. The loss estimate includes two types of magstripe "discretionary" data that was not being properly verified: the CVV and an optional PIN offset used by some banks.

Chapter 16: Operation Firewall

105 *Banner ads appeared at the top of the site:* This and other reporting on Shadowcrew's contents comes from a mirror of the public portion of the site captured in October 2004, immediately before it was shuttered.

107 *The posts disappeared at once:* Interviews with Max. Aragon independently stated that he and Max tried to warn Shadowcrew members in advance of the Operation Firewall raids.

108 *The transactions ranged from the petty to the gargantuan:* Transaction details come from the Operation Firewall indictment, *U.S. v. Mantovani et al.,* 2:04-cr-00786, U.S. District Court for the District of New Jersey.

109 *the Secret Service had noticed Ethics was selling:* Ethics's hacking of the Secret

Service agent was first reported by the author: "Hacker penetrates T-Mobile systems," Securityfocus.com, January 11, 2005. His use of the BEA Systems exploit came from sources close to the case and was first reported by the author: "Known Hole Aided T-Mobile Breach," Wired.com, February 28, 2005 (http://www.wired.com/politics/security/news/2005/02/66735). Also see *U.S. v. Nicolas Lee Jacobsen,* 2:04-mj-02550, U.S. District Court for the Central District of California.

110 *David Thomas was a lifelong scammer who'd discovered the crime forums:* For Thomas's history with the forums and the details of his work for the FBI, see Kim Zetter, "I Was a Cybercrook for the FBI," Wired.com, January 20, 2007. A U.S. government source confirmed to the author that Thomas had worked for the bureau while running his forum, the Grifters.

111 *"You don't know who you have here":* From the police report of Thomas's arrest. "The problem with the Bureau and the Secret Service is they look at the largest biggest deals they can get in on," Thomas said in a 2005 interview with the author. "They want the big enchilada."

113 *Their targets were marked on a map of the United States:* Brian Grow, "Hacker Hunters," *Businessweek,* May 30, 2005 (http://www.businessweek.com/magazine/content/05_22/b3935001_mz001.htm). The identification of the Secret Service agents' guns also comes from this story.

113 *Attorney General John Ashcroft boasted in a press release: "Nineteen Individuals Indicted in Internet 'Carding' Conspiracy,"* October 28, 2004 (http://www.justice.gov/usao/nj/press/files/pdffiles/fire1028rel.pdf).

Chapter 17: Pizza and Plastic

116 *His scanning put him inside a Windows machine:* Max, Jonathan Giannone, and Brett Johnson each independently identified the Pizza Schmizza in Vancouver, Washington, as the source of Max's dumps in this period. The store manager said the restaurant has since changed ownership, and she had no knowledge of a breach.

117 *Max couldn't help feeling cheated yet again:* Interviews with Max.

117 *Giannone was a smart middle-class kid with a coke habit:* Giannone confirmed the cocaine use and all the details of his relationship with Max and Aragon. He discussed the elevator button pressing and the "bank robbery" prank in a chat with another carder, a log of which was provided to the author. Giannone confirmed in an in-

terview that he discussed the bank robbery hoax but said it was an idle boast, and he didn't actually pull it off. He said he did not recall the elevator matter.

118 *Giannone joined Shadowcrew and CarderPlanet under the handle MarkRich:* Giannone's transition through various handles was confirmed by Giannone in an interview. Posts on the forums reviewed by the author confirm he gave up his original handle after being suspected of informing on an associate while a juvenile.

118 *launched a DDoS attack against JetBlue:* Giannone also discussed this attack in the abovementioned chat logs. He confirmed it in interviews with the author.

118 *the teen was running his operations from the computer in his mother's bedroom:* Interviews with Max.

Chapter 18: The Briefing

120 *Mularski had wanted to be an FBI agent since his freshman year:* Mularski's biographical details and his early work at NCFTA come from interviews with Mularski.

122 *The briefing for about half a dozen FBI agents:* Interviews with J. Keith Mularski and Postal Inspector Greg Crabb.

Chapter 19: Carders Market

124 *"Sherwood Forest" wasn't going to cut it for a criminal marketplace:* Aragon's rejection of the name comes from interviews with Max and a letter Max later wrote his sentencing judge.

125 *Janer, an avid watch collector, headed straight to Richard's:* Janer explained his motives in the failed watch caper in interviews, and Aragon confirmed he provided Janer with cards as a favor. The criminal case file describes how he was busted and his subsequent cooperation, which Janer confirmed. *U.S. v. Werner William Janer,* 3:06-cr-00003, U.S. District Court for the District of Connecticut.

127 *He hacked into a Florida data center run by Affinity Internet:* Court records confirm Carders Market was hosted at Affinity at this time and that Affinity later provided the FBI with a copy of the file system. Max detailed the hack in interviews and in contemporaneous postings to an Internet message board as "Iceman."

128 *"I'm looking to make a good pile of money":* Chat logs admitted as evidence in *U.S. v. Jonathan Giannone,* 3:06-cr-01011, U.S. District Court for the District of South Carolina. Online chats and message board posts in this book are verbatim when they appear within quotes, except for some minor changes of grammar, punctuation, or spelling for readability.

Chapter 20: The Starlight Room

130 *Tsengeltsetseg Tsetsendelger was being kissed:* Aragon, Max, and other sources confirm that Tsetsendelger was recruited at the Starlight Room and brought back to Aragon's hotel. The details come from interviews with Tsetsendelger. Liz and Michelle Esquere declined comment.

Chapter 22: Enemies

140 *required technicians to reboot the machine every 49.7 days:* Sources include Linda Geppert, "Lost Radio Contact Leaves Pilots on Their Own," *IEEE Spectrum,* November 2004 (http://spectrum.ieee.org/aerospace/aviation/lost-radio-contact-leaves-pilots-on-their-own).

141 *Giannone was pretty sure he couldn't hack Macs:* Interview with Giannone. Max acknowledges that he hacked Giannone frequently and tracked his movements, and was also prone to sending long messages to Giannone, and others, reflecting his thoughts. He also clarified that he had no problem hacking Macs.

142 *So he reached out to Thomas by ICQ to try to head off trouble:* Max and Aragon discussed their ongoing conflict with Thomas, who also detailed his suspicions about Carders Market and Johnson on his own website, the Grifters. Additionally, the author obtained a log of the chat between Aragon and Thomas quoted herein.

Chapter 23: Anglerphish

146 *He needed the money, plain and simple:* Johnson's personal story comes from a sworn affidavit he filed in his criminal case on April 13, 2007, and a letter he wrote his sentencing judge on March 1, 2007. See *U.S. v. Brett Shannon Johnson,* 3:06-cr-01129, U.S. District Court for the District of South Carolina.

146 *displayed simultaneously on a forty-two-inch plasma screen hanging on the wall of the office:* Trial transcript in *U.S. v. Jonathan Giannone,* 3:06-cr-01011, U.S. District Court for the District of South Carolina.

147 *The suspect had done everything but deep-clean the carpet and paint the walls:* Interview with Justin Feffer, senior investigator, High Technology Crime Division, Los Angeles County District Attorney's Office. Also see *The People of the State of California v. Shawn Mimbs,* BA300469, Superior Court of California, County of Los Angeles. Mimbs declined comment.

147 *The needles were steady as Johnson answered the first two questions:* According to Johnson. The Secret Service declined to discuss Operation Anglerphish.

148 *"I will hound you for the rest of your life":* From Johnson's letter to his sentencing judge.

Chapter 24: Exposure

150 *"Tea, these girls are white trash":* Interview with Tsengeltsetseg Tsetsendelger. Aragon mentioned his fondness for Tsetsendelger in interviews and a letter to the author.

151 *Iceman, she'd decided, was pretty cool:* Interview with Tsetsendelger. Max says he was respectful in chats with her but privately disliked her.

152 *"Get out of here":* The incident at the pool comes from interviews with Tsetsendelger and Giannone.

153 *The bug was in the brief handshake sequence:* See CERT Vulnerability Note VU#117929. The bug was discovered accidentally by Steve Wiseman of Intelliadmin.com while he was writing and testing a VNC client. Technical details come from an analysis by James Evans; see http://marc.info/?l=bugtraq&m=114771408013890&w=2.

156 *a widely read computer security blog:* "Schneier on Security" by Bruce Schneier. http://www.schneier.com/blog/archives/2006/06/interview_with_1.html.

157 *a random blog called "Life on the Road":* See http://afterlife.wordpress.com/2006/06/19/cardersmarket-shadowcrew-and-credit-card-theft/ and http://afterlife.wordpress.com/2006/07/12/carding-web-sites/.

Chapter 25: Hostile Takeover

162 *Carders Market had six thousand members now:* Max, his former administrator Th3C0rrupted0ne, and other carders say the site had in excess of six thousand users after the hostile takeover. The Justice Department, though, has put the number at forty-five hundred.

166 *secret even from his mother:* According to his mother, Marlene Aragon.

Chapter 26: What's in Your Wallet?

171 *industry-funded report by Javelin Research:* Javelin Strategy and Research, "2007 Identity Fraud Survey Report," February 2007. The report was sponsored by Visa USA, Wells Fargo, and CheckFree, and then prominently cited by Visa USA in a PowerPoint presentation at a Federal Trade Commission workshop: "50% of known thieves—*were known by the victim!*" (emphasis original). Also see the author's "Stolen Wallets, Not Hacks, Cause the Most ID Theft? Debunked," Wired

.com, February 12, 2009 (http://www.wired.com/threatlevel/2009/02/stolen-wallets/).

171 *Visa's private numbers told the real story:* Presentation by Steven Johnson, director, Visa U.S.A. Public Sector Sales, at the ninth annual GSA SmartPay Conference in Philadelphia, August 23, 2007. The presentation slides are marked "Visa Confidential."

172 *C0rrupted had discovered the warez scene on dial-up bulletin board systems:* Biographical information comes from telephone and online interviews with Th3C0rrupted0ne, who spoke on condition that his real name not be reported.

173 *"I can't believe how much you know about me":* Interview with Aragon.

174 *"Do not follow unsolicited links":* US-CERT Technical Cyber Security Alert TA06-262A (http://www.kb.cert.org/vuls/id/416092).

175 *Each copy of the message was customized:* The text of the spear phishing e-mail comes from an FBI affidavit filed in *U.S. v. Max Ray Butler,* 3:07-mj-00438, U.S. District Court for the Eastern District of Virginia. "Mary Rheingold" is not a real name and was added by the author in place of "[First Name and Last Name of Recipient]" in the original court document.

Chapter 27: Web War One

180 *"The Secret Service and FBI declined to comment on Iceman or the takeovers":* Byron Acohido and Jon Swartz, "Cybercrime flourishes in online hacker forums," *USA Today,* October 11, 2006.

180 *"You've lost your fucking mind":* Interview with Chris Aragon.

180 *Bank of America and Capital One, in particular, were huge institutions:* Of his spear-phishing attacks, Max was charged only with the Capital One intrusion. The other victims were identified by Max.

Chapter 28: Carder Court

184 *it was just Silo trying to gather intelligence on DarkMarket members for the police:* Max, Mularski, and Th3C0rrupted0ne identified Liske as Silo. In extensive interviews, Liske was evasive about his activities on the forums but spoke obliquely of his work as an informant and his relationship with Max. "Max was a good case. You know, he was a challenge." On the NCFTA Trojan, he said: "Isn't it reasonable to assume that whoever was dishing out Trojans was actually dishing out Trojans to everyone in the scene?" Later, "If it were malicious I could have—someone could have caused real damage." Detective Mark Fenton of the Vancouver Police

Department said Canadian law prohibits him from identifying or confirming an informant's identity. On the subject of whether he received hacked evidence from informants, he said: "I know down in the States, if an individual received any information that is suspect, it's not admissible. Up here, if someone tells me something, I say, 'Where did you hear that from?' He says, 'I heard it from some guy.'" He likened the arrangement to the Crime Stoppers tip program. "Should Crime Stoppers be scrapped because we have criminals phoning in tips about other criminals?" One unanswered question is to what degree, if any, the Secret Service leaned on hacked information provided by the VPD to build cases in the United States. The Secret Service declined to make agents available to the author: "Although we have chosen not [to] participate with this particular project, feel free to approach us with other ideas in the future."

188 *the same user had once registered another address through the company:* Max says Night Fox was responsible for registering the Financial Edge News website and made this blunder.

Chapter 29: One Plat and Six Classics

192 *"for 150 classics":* Affidavit of Secret Service Special Agent Roy Dotson, July 24, 2007, filed in *USA v. E-Gold, LTD,* 1:07-cr-0019, U.S. District Court for the District of Columbia. For the complete history of e-gold, see Kim Zetter, "Bullion and Bandits: The Improbable Rise and Fall of E-Gold," Wired.com, June 9, 2007.

193 *They were working closely with Silo's handler at the Vancouver Police Department:* Word of the meeting got back to Liske. "There was an accusation that I was Iceman," he said in an interview. "And there was a big presentation made that this guy was Iceman. And the people this was presented to knew full well that I wasn't."

Chapter 30: Maksik

195 *straight from Maksik's massive database of stolen cards: U.S. v. Maksym Yastremski,* 3:06-cr-01989, U.S. District Court for the Southern District of California.

197 *In early 2006, the Ukranians finally identified Maksik as one Maksym Yastremski:* Interview with Greg Crabb.

198 *they secretly copied his hard drive for analysis:* Government filing dated July 24, 2009, in *U.S. v. Albert Gonzalez,* 2:08-cr-00160, U.S. District Court for the Eastern District of New York.

199 *"We were lucky in this case, because Salgado's purchaser was cooperating with the FBI":*

Written testimony of Robert S. Litt, deputy attorney general, before the Subcommittee on Telecommunications, Trade and Consumer Protection, House Commerce Committee, September 4, 1997 (http://www.justice.gov/criminal/cybercrime/daag9_97.htm).

199 *But the feds lost the crypto wars:* For a detailed history, see Steven Levy, *Crypto: How the Code Rebels Beat the Government—Saving Privacy in the Digital Age* (New York: Penguin Books, 2002).

Chapter 31: The Trial

202 *"So, you take my girls out to party now?":* Interview with Giannone.

202 *Once a jury is seated, a defendant's chances for acquittal are about one in ten:* Fiscal year 2006. Calculated from "Federal Justice Statistics, 2006—Statistical Tables," U.S. Department of Justice, Bureau of Justice Statistics, May 1, 2009 (http://bjs.ojp.usdoj.gov/index.cfm?ty=pbdetail&iid=980).

203 *"I suspect that you are never going to look at the Internet exactly the same way again":* Trial transcript in *U.S. v. Jonathan Giannone,* 3:06-cr-01011, U.S. District Court for the District of South Carolina. Some grammatical changes were made for readability.

204 *"Who's Iceman?":* Interview with Giannone.

Chapter 32: The Mall

208 *his new partner, twenty-three-year-old Guy Shitrit:* Information about Shitrit's trouble in Miami comes from Aragon. Detective Robert Watts of the Newport Beach Police Department confirmed he'd heard the same account. Shitrit, now in custody, did not respond to a letter from the author.

208 *His wife, Clara, had brought in $780,000 on eBay in a little over three years:* Based on sales figures from Clara Aragon's eBay account obtained by the Newport Beach Police Department. Aragon declined to discuss his profits.

208 *Max, he felt, was ignoring the Whiz List, their blueprint for building one big score and getting out:* Interview with Aragon. When police searched Aragon's cell phone, they found this entry on his electronic to-do list: "tackle whiz list."

209 *in meticulous, hand-drawn spreadsheets summing up how much Chris owed her for each in-store appearance:* One such spreadsheet was seized by the Newport Beach Police Department and seen by the author.

209 *Vigo was looking for a way to pay down a $100,000 debt to the Mexican Mafia:* This

according to Vigo's statements to the police following his arrest. The Newport Beach Police Department found a copy of the shipping manifest in Vigo's office.

211 *Bloomingdale's security people didn't like to upset the store's customers:* Interview with Detective Robert Watts.

211 *thirty-one Coach bags, twelve new Canon PowerShot digital cameras:* Per the search warrant seizure records.

Chapter 33: Exit Strategy

214 *Max decided to invest in a rope ladder:* Interview with Max.

214 *Max finally learned about Giannone's bust from a news article:* Kim Zetter, "Secret Service Operative Moonlights as Identity Thief," Wired.com. June 6, 2007 (http://www.wired.com/politics/law/news/2007/06/secret_service).

215 *He was growing jumpier every day:* Based on an interview with Charity Majors. Max says he was alert but not jumpy.

215 *a judge approved his legal name change from Max Butler to Max Ray Vision:* In *Re: Max Ray Butler,* CNC-07-543988, County of San Francisco, Superior Court of California.

216 *Silo had hidden a second message:* Interview with Max. Lloyd Liske would neither confirm nor deny this account.

217 *The company openly marketed the service as a way to circumvent FBI surveillance:* "In some countries, government sponsored projects have been set up to collect massive amounts of data from the Internet, including emails, and store them away for future analysis. [. . .] One example of such a program was the FBI's Carnivore project. By using Hushmail, you can be assured that your data will be protected from that kind of broad government surveillance." http://www.hushmail.com/about/technology/security/.

217 *forced Hushmail officials to sabotage their own system and compromise specific surveillance targets' decryption keys:* Ryan Singel, "Encrypted E-Mail Company Hushmail Spills to Feds," Wired.com. November 7, 2007. Detective Mark Fenton of the Vancouver Police Department said he provided Max's Hushmail e-mail to the Secret Service.

217 *It was supposed to be a training run for one of Chris's new recruits:* Interviews with Tsengeltsetseg Tsetsendelger and Chris Aragon.

218 *a female Secret Service agent disguised as a maid:* The Secret Service's surveillance,

including the ride up the elevator with Max, was described in an affidavit in *U.S. v. Max Ray Butler,* 2:07-cr-00332, U.S. District Court for the Western District of Pennsylvania. Max said in an interview that the agent was dressed as a maid. FBI agent Mularski says the surveillance was on and off for months.

218 *Chris picked out Max's mugshot from the photos: U.S. v. Max Ray Butler,* 2:07-cr-00332, U.S. District Court for the Western District of Pennsylvania. Aragon says the government tricked him by telling him Max had already been arrested, but he also gave them information on Max's security measures, which undermines that claim. Court records for Aragon's criminal case in Orange County indicate a sealed letter from Dembosky is on file. *The People of the State of California vs. Christopher John Aragon, et al.,* 07HF0992, Superior Court of California, County of Orange.

220 *Two had lost power when an agent tripped over an electrical cable:* According to Max.

220 *Max's head snapped to look at Master Splyntr:* Interview with Mularski.

221 *"You were right":* Interview with Charity Majors.

221 *"Why do you hate us?":* Interview with Max.

Chapter 34: DarkMarket

225 *he told a harrowing story:* "Son bilgiyi verecekken yok oldu!" Haber 71, August 12, 2008 (http://www.haber7.com/haber/20080812/Son-bilgiyi-verecekken-yok-oldu.php).

226 *fingering a known member of Cha0's organization as the shipper:* Mularski described the genesis of the investigation. The role played by the shipping companies was detailed by Uri Rivner of RSA in a blog post (http://www.rsa.com/blog/blog_entry.aspx?id=1451). The Turkish National Police referred inquiries to their embassy in Washington, DC, which declined to make detectives available for interviews.

226 *a tall, beefy man with close-cropped hair and a black T-shirt emblazoned with the Grim Reaper:* Per police video of the arrest and search. Also see "Enselenen Chao sanal şemayı anlattı," Haber 7, September 12, 2008 (http://www.haber7.com/haber/20080912/Enselenen-Chao-sanal-semayi-anlatti.php).

226 *matching his appearances at the Java Bean with JiLsi's posts:* Interview with Mularski. Also see Caroline Davies, "Welcome to DarkMarket—global one-stop shop for cybercrime and banking fraud," *Guardian,* January 4, 2010 (http://www.guardian.co.uk/technology/2010/jan/14/darkmarket-online-fraud-trial-wembley).

226 *JiLsi's associate, sixty-seven-year-old John "Devilman" McHugh,* Ibid.

227 *Erkan "Seagate" Findikoglu:* Interview with Mularski. Also see Fusun S. Nebil, "FBI Siber Suçlarla, ABD İçinde ve Dışında İşbirlikleri ile Mücadele," Turk .internet.com, June 15, 2010 (http://www.turk.internet.com/portal/yazigoster .php?yaziid=28171).

227 *Twenty-seven members of Seagate's organization were charged in Turkey:* Interview with Mularski.

228 *a reporter for Südwestrundfunk, Southwest Germany public radio:* The reporter was Kai Laufen. See http://www.swr.de/swr2/programm/sendungen/wissen/-/id=660374/nid=660374/did=3904422/p6601i/index.html.

228 *The U.S. press picked up the story:* The author was the first to identify J. Keith Mularski by name as the FBI agent posing as Master Splyntr, in "Cybercrime Supersite 'DarkMarket' Was FBI Sting, Documents Confirm," Wired.com, October 13, 2008 (http://www.wired.com/threatlevel/2008/10/darkmarket-post/).

Chapter 35: Sentencing

229 *It had taken the CERT investigators only two weeks to find the encryption key:* Max well knew that the key was vulnerable while in RAM, but he believed the software security on his server would prevent anyone from gaining access to its memory. CERT's Matt Geiger, who led the forensics team, declined to comment on how he bypassed that security but he said he was able to run memory-acquisition software on Max's computer.

230 *Max had stolen 1.1 million of the cards from point-of-sale systems:* Max didn't challenge this amount for sentencing, but in interviews he expressed disbelief that the number could be that high.

Chapter 36: Aftermath

234 *An undercover Secret Service operative lured him to a nightclub: "2010 Data Breach Investigations Report,"* Verizon RISK Team in cooperation with the United States Secret Service, July 28, 2010.

234 *ICQ user 201679996:* Affidavit In Support of Arrest Warrant, May 8, 2007, *U.S. v. Albert Gonzalez,* 2:08-mj-00444, U.S. District Court for the Eastern District of New York.

236 *it was Jonathan James who would pay the highest price:* See the author's "Former Teen Hacker's Suicide Linked to TJX Probe," Wired.com, July 9, 2009 (http://www .wired.com/threatlevel/2009/07/hacker/).

237 *They recruit ordinary consumers as unwitting money launderers:* For more detail on these so-called "money mule" scams, see the blog of former Washingtonpost .com reporter Brian Krebs, who has covered the crime extensively: http:// krebsonsecurity.com/.

238 *the Secret Service had been paying Gonzalez an annual salary of $75,000 a year:* First reported in Kim Zetter, "Secret Service Paid TJX Hacker $75,000 a Year," Wired .com, March 22, 2010.

238 *filed by the attorneys general of 41 states:* Sources include Dan Kaplan, "TJX settles over breach with 41 states for $9.75 million," *SC Magazine,* June 23, 2009 (http://www.scmagazineus.com/tjx-settles-over-breach-with-41-states-for-975-million/article/138930/).

238 *another $40 million to Visa-issuing banks:* Mark Jewell, "TJX to pay up to $40.9 million in settlement with Visa over data breach," Associated Press, November 30, 2007.

238 *Heartland had been certified PCI compliant:* Sources include Ellen Messmer, "Heartland breach raises questions about PCI standard's effectiveness," *Network World,* January 22, 2009 (http://www.networkworld.com/news/2009/012209-heartland-breach.html).

238 *Hannaford Brothers won the security certification even as hackers were in its systems:* Sources include Andrew Conry-Murray, "Supermarket Breach Calls PCI Compliance into Question," *InformationWeek,* March 22, 2008.

238 *The restaurants filed a class-action lawsuit:* http://www.prlog.org/10425165-secret-service-investigation-lawsuit-cast-shadow-over-radiant-systems-and-distributo .html. Also, "Radiant Systems and Computer World responsible for breach affecting restaurants—lawsuit," Databreaches.net, November 24, 2010 (http://www .databreaches.net/?p=8408) and Kim Zetter, "Restaurants Sue Vendor for Unsecured Card Processor," Wired.com, November 30, 2009 (http://www.wired .com/threatlevel/2009/11/pos).

239 *White hats have devised attacks against chip-and-PIN:* See Steven J. Murdoch, Saar Drimer, Ross Anderson, and Mike Bond, "Chip and PIN Is Broken," University of Cambridge Computer Laboratory, Cambridge, UK. Presented at the 2010 IEEE Symposium on Security and Privacy, May 2010 (http://www.cl .cam.ac.uk/research/security/banking/nopin/). The response by the UK Card Association is at http://www.theukcardsassociation.org.uk/view_point_and_ publications/what_we_think/-/page/906/.

239 *hundreds of thousands of point-of-sale terminals with new gear:* The cards themselves are more expensive as well. For a more thorough discussion of the issues holding back chip-and-PIN's adoption in the United States, see Clases Bell, "Are chip and PIN credit cards coming?" Bankrate.com, February 18, 2010 (http://www.foxbusiness.com/story/personal-finance/financial-planning/chip-pin-credit-cards-coming/). See also Allie Johnson, "U.S. credit cards becoming outdated, less usable abroad," Creditcards.com (http://www.creditcards.com/credit-card-news/outdated-smart-card-chip-pin-1273.php).

Epilogue

241 *His mother suggested he get an agent:* A letter to the author from Aragon.

ACKNOWLEDGMENTS

I first encountered Max Vision some ten years ago, when I was a newbie reporter for the computer security site SecurityFocus.com. Max was then facing charges over his scripted attack on thousands of Pentagon systems, and I was fascinated by the story playing out in the Silicon Valley courtroom, where the federal justice system was bearing down on a once-respected computer security expert who'd upended his life with a single, quixotic hack.

Years later, after I'd reported on hundreds of computer crimes, vulnerabilities, and software glitches, Max was arrested again, and a new federal indictment exposed the secret life he'd led after his fall from grace. As I investigated, I grew certain that Max, more than anyone else, embodied the sea change I'd witnessed in the world of hacking, and would be the perfect lens through which to explore the modern computer underground.

Fortunately, others agreed. I owe a debt of thanks to my agent, David Fugate, who guided me through the process of developing my idea into a book proposal, and my editor at Crown, Julian Pavia, who worked tirelessly to keep me on course and only slightly behind schedule throughout a year of reporting, writing, and rewriting.

Also crucial was the enormous support from my boss, Evan Hansen, editor in chief at Wired.com. And I'm grateful to my colleagues at Wired.com's Threat Level blog, Kim Zetter, Ryan Singel, and David Kravets, who collectively shouldered the burden of my absence for two months while I finished the book and then braved the burden of my irritable, bleary-eyed return afterward.

My thanks also to Joel Deane and Todd Lapin, who showed me the ropes when I became a journalist in 1998, and Al Huger and Dean Turner of SecurityFocus.com. Jason Tanz at *Wired* magazine did an amazing job with my feature article on Max, "Catch Me If You Can," in the January 2009 issue.

Among my guides in this book were the cops, feds, hackers, and carders who spoke with me at length, with no benefit to themselves. FBI Supervisory Special Agent J. Keith Mularski was particularly generous with his time, and Max Vision spent many hours on the prison phone and writing long e-mails and letters to share his story with me.

My thanks to U.S. Postal Inspector Greg Crabb, Detective Bob Watts of the Newport Beach Police Department, former FBI agent E. J. Hilbert, and Assistant U.S. Attorney Luke Dembosky, the latter of whom wouldn't tell me much, but was always nice about it. And I'm grateful to Lord Cyric, Lloyd Liske, Th3C0rrupted0ne, Chris Aragon, Jonathan Giannone, Tsengeltsetseg Tsetsendelger, Werner Janer, Cesar Carranza, and other veterans of the carder scene who asked to remain unnamed.

The story of Max Vision would have listed heavily to his criminal side were it not for Tim Spencer and Marty Roesch, who shared their experience of Max as white-hat hacker, and Kimi Mack, who spoke candidly about her marriage to Max. My thanks also to security wunderkind Marc Maiffret, who helped isolate some of Max's exploits.

The underworld that *Kingpin* delves into has been illuminated by a number of first-rate journalists, including Bob Sullivan, Brian Krebs, Joseph Menn, Byron Acohido, Jon Swartz, and my *Wired* colleague Kim Zetter.

Finally, my thanks to my wife, Lauren Gelman, without whose loving support and sacrifice this book would not have been possible, and to Sadelle and Asher, who will find their computer use closely supervised until they're eighteen.

ABOUT THE AUTHOR

KEVIN POULSEN is a senior editor at Wired.com and a contributor to *Wired* magazine. He oversees cybercrime, privacy, and political coverage for Wired.com and edits the award-winning Threat Level blog (wired.com/threatlevel), which he founded in 2005. He's broken numerous national stories, including the FBI's use of spyware in criminal and national security investigations; a hacker's penetration of a Secret Service agent's confidential files; and the secret arrest of an Army intelligence officer accused of leaking documents to whistle-blowing website WikiLeaks. In 2009 he was inducted into MIN's Digital Hall of Fame for online journalism and in 2010 was voted one of the "Top Cyber Security Journalists" by his peers.